Principles of Organic Chemistry

The Structure of the Atom

by

Stephen John Shroyer

Preface

The Front matter to organic chemistry is often not described because it is learned in earlier subjects. So it is quickly reviewed, assuming you already know the matter.

This text correlates the physics and math as it relates to organic chemistry.

The usual organic chemistry course spends very little time on this material.

By studying the details on this matter, the understanding of organic chemistry becomes a tenable subject.

It answers a lot of the "why"?

It goes from memorization to knowledge.

Topics

Mendeleev's Periodic Table
Bohr's Planetary Model
Orbits are quantized, energy related.
DeBroglie's Duality
Heisenberg's Uncertainty
Quantum: Wave Model
Schrodinger's Wave Equation
Wave Packet is the particle
Psi

<x>

|psi|^2 : Probability Density
Quantum numbers: n, l, ml, ms
Allowed states of the atom
Normalization
Atomic Orbits
Pauli Exclusion Principle
Aufbau Principle
Hund's Rule

Electron Configuration
Orbital Energy Diagram
Magnetism
Atomic Radii
Ionization Energy
Electron Affinity
Flame colors
Oxidation
Reduction
Acid-Base
Lewis: Octet Rule
Electronegativity
Molecular Orbitals
Orbital Hybridization
VSEPR, VBT, MOT
Dipole
Polarity
Induction
Field Effects
Formal Charge
Resonance

Structure

Atom

Dmitri Ivanovich Mendeleev

Russia 1871

The Periodic Table

The element is the simplest and purest substance made of the same atom.

Similar physical and chemical properties (Solid, liquid, gas and unknown).

7 rows = periods

18 columns=groups (similar properties which occur periodically)=families.   Group numbers: A1 → 8 and B1 → 8.

Atomic weight and number increases Left → Right for the most part.

Metals on the Left. Nonmetals on the Right. Metalloids, semimetals, are in between them.

1	Atomic number=#p+ = Z
H	Symbol
Hydrogen	Name
1.00794	Atomic mass (weighted average)

In organic chemistry, we are mainly interested in H,C,N and O.

1913
A Planetary Model of the Atom

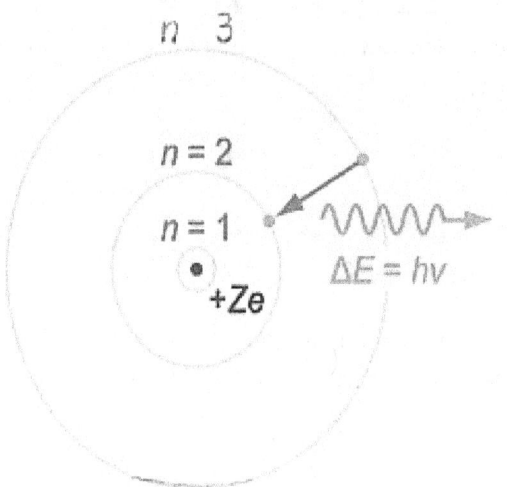

The **Rutherford–Bohr model** of the <u>hydrogen atom</u> ($Z = 1$) or a hydrogen-like ion ($Z > 1$), where the negatively charged <u>electron</u> confined to an <u>atomic shell</u> encircles a small, positively charged <u>atomic nucleus</u> and where an electron jump between orbits is accompanied by an emitted or absorbed amount of <u>electromagnetic energy</u> ($h\nu$).[1] The orbits in which the electron may travel are shown as grey circles; their radius increases as n^2, where n is the <u>principal quantum number</u>. The 3 → 2 transition depicted here produces the first line of the <u>Balmer series</u>, and for hydrogen ($Z = 1$) it results in a photon of <u>wavelength</u> 656 <u>nm</u> (red light).

This model is accurate for a single electron atom type, eg. H, He+, and Li2+.

The nucleus is occupied by the single proton and neutron. The electron orbits around the nucleus, like a planet around the sun. Electrons tend to stay in the lowest energy levels; the ones closest to the nucleus. This is called the ground state. When an energy source raises the electron to a higher energy level, the electron is in the excited state. When the electron returns to the ground state, energy is released as light. The more levels dropped, the larger the frequency and the shorter the wavelength. A particle moving in circular motion has angular momentum. Momentum is the mass times the velocity. Both the momentum and energy level of the atom is quantized. This means that the values of both are discrete, not infinite or changing. This enables the electron to stay in orbit without crashing into the nucleus. For energy, the following relationship was given:

$$E(n) = -B/n^2$$

E=energy
n=integer, # of orbit
B=$2.179*10^{-18}$J J=Joules
(-) sign = attraction
E=0 electron at infinite distance from

nucleus.

B is a constant. n is the variable of the function. As n increases the attractive force with said amount of energy decreases. This makes the electron able to leave orbit and get involved in chemical reactions. In an excited state, the electron is able to participate in a chemical reaction. It carries with it energy to do so. The electron is prepared to ionize (dissociate) from the nucleus.

Notes:

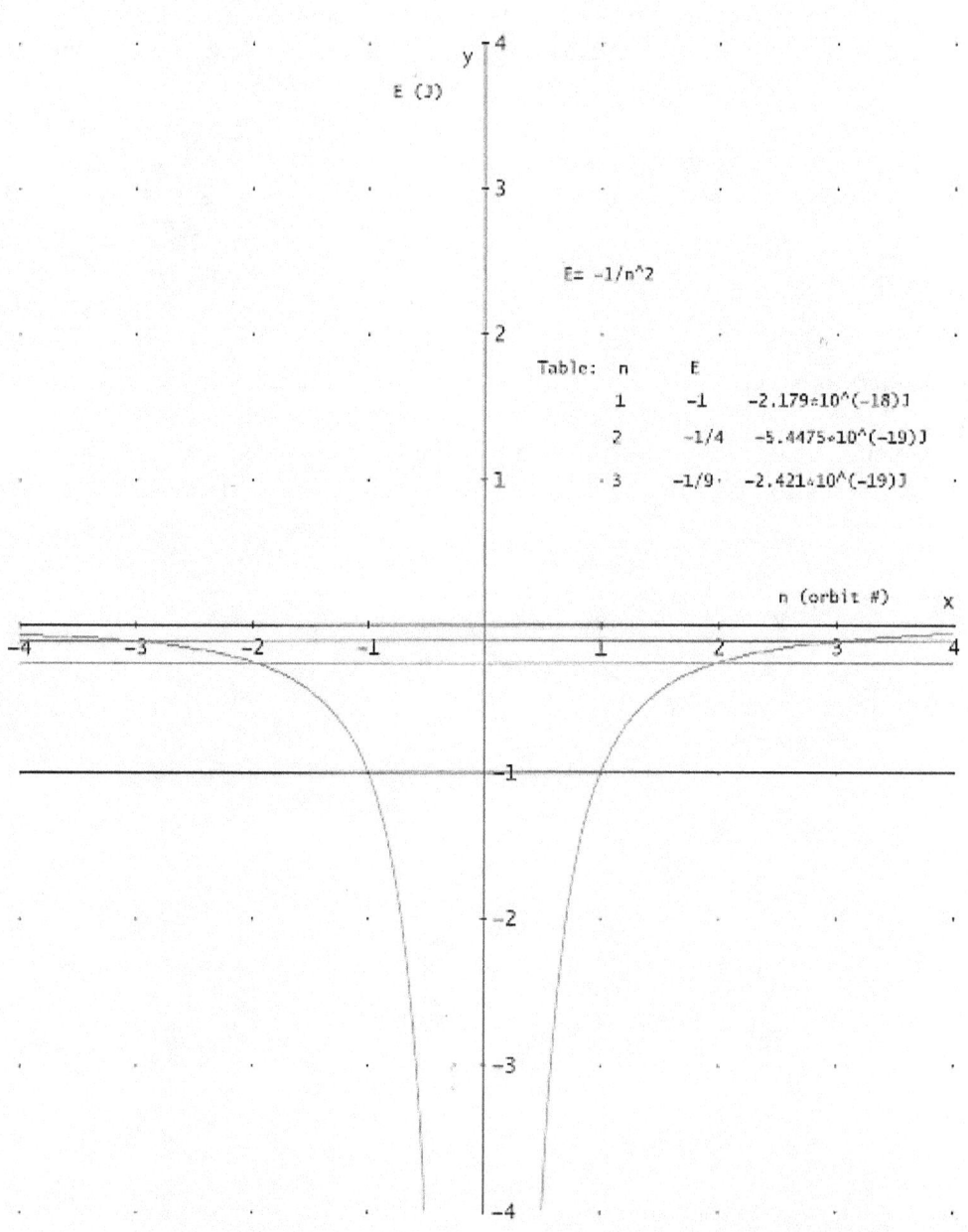

E= -1/n^2

Table: n   E
       1   -1   -2.179÷10^(-18)J
       2   -1/4   -5.4475÷10^(-19)J
       3   -1/9   -2.421÷10^(-19)J

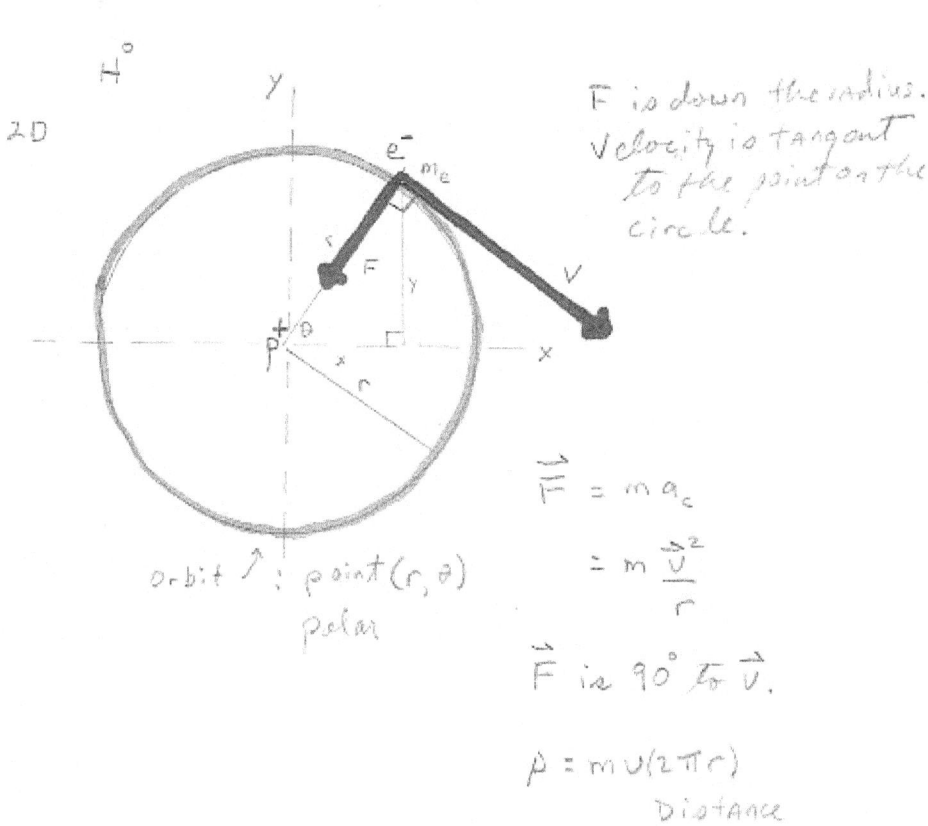

The geometry explains the idea! The motion is circular. In reality, it is 3D. But we only need 2D for the analysis.

The F vector is radial towards the center, because of the electrical attractive force between p+ and e-. The sign is (-) because of (-1) * (+1) = -1.

The velocity vector is tangent to the circle and perpendicular to the force vector.

Newtonian mechanics explains all of this.

Here are the new ideas:

The preferred orbit is the one with the least amount of energy, which is not changing. The orbit is maintained and is discrete.

If there is a change in energy by changing the orbit, the change is quantized. The orbit is chosen by an integer multiple of Planck's constant.

$$r \text{ proportional to } n * h$$
radius   integer   Planck's constant

The force, mass, velocity, acceleration is constant.

The radius and energy is changing and it is quantized: multiple of n.

Bohr's model of the H atom can be used to analyze quantized energy and angular momentum of atom-sized systems and other properties of the H atom.

Here are his ideas:
1. Electron in circular orbit under the influence of a electric force
                                of attraction.
  (-)charge   2π=360°   Coulomb's law
                                            (similar to Gravitation Law)
2. In a stable orbit, there is no loss of energy. Therefore the electron
maintains it's orbit.   This is the default orbit.
The total energy of the atom is constant. Lowest energy=stable.

3. During transition, energy is absorbed or emitted as a photon whose
                                            f α ΔE not motion.
Δ orbit   quantized energy   E(i)-E(f) = hf i   > f
                                                 ΔE   initial final
f=frequency (1/s) E=energy Δ=change h=Planck's constant
4. Size of orbit α quantized angular momentum: m(e)vr=nh(bar)
h(bar) =h/2π n=1,2,3...   n=integer
                                        m(e)v(2πr)=nh
                              Circumference=distance around a circle
                         momentum=mass * velocity * distance=force
m(e)=mass of electron v=velocity r=radius π=pi constant
momentum in circular motion is equated to planck's contant in circular
dimension times an integer multiple.
                        digitilized

$\Delta E(J)=h(J*s)*f \rightarrow f=1/s$

#1: $e = h \cdot f$

#2: $e = \dfrac{3313 \cdot f}{500000000000000000000000000000000000}$

#3: $e = 6.626 \cdot 10^{-34} \cdot f$

Planck's constant

#4: $\dfrac{h}{2 \cdot \pi}$

#5: $\dfrac{3313}{1000000000000000000000000000000000000 \cdot \pi}$

#6: $1.054560652 \cdot 10^{-34}$

h(bar)

The potenital energy of the system:
$U = (k(e)q_1q_2)/r \rightarrow - (k(e) e^2)/r$
   Coulomb constant $= 8.987*10^9$ (N*m^2)/C^2
  $1/(4\pi\varepsilon(0))$
       $8.854*10^{-12}$ C^2/(N*m^2)
       permittivity of free space: Electric density and field
       $1/\mu(0)*c^2$
           speed of light $= 3*10^8$ m/s
      $4\pi*10^{-7}$T*m/A T=Tesla
      permeability of free space: Magnetic density and field

Total energy of the system:
$E = K$ $+U$
  $1/2$ m(e)v^2 $-k(e)e^2/r$
  Kinetic energy   Potential energy
  energy of motion energy of position

F   =   m   a Newton's 2nd law
$k(e)e^2/r^2$ m(e) v^2/r

$K = 1/2$ m(e)v^2 = k(e) e^2/2r $mv^2 = k e^2/r \rightarrow 1/2 mv^2 = k e^2/2r$

$E = -k e^2/2r$ $E = 1/2r - 1/r \rightarrow -1/2r$
The amount of energy to be added to the symtem to remove the electron. E->0.

Notes:

Solve the quantized momentum equation for v^2 and the Kinetic energy equation
for v^2. These are now both equal to v^2. Now solve for radius of the
orbit.

#1: m·v·r = n·h

#2: SOLVE(m·v·r = n·h, v)

#3:
$$v = \frac{h \cdot n}{m \cdot r}$$

#4:
$$v^2 = \left(\frac{h \cdot n}{m \cdot r}\right)^2$$

#5:
$$\frac{1}{2} \cdot m \cdot v^2 = \frac{k \cdot e^2}{2 \cdot r}$$

#6:
$$SOLVE\left(\frac{1}{2} \cdot m \cdot v^2 = \frac{k \cdot e^2}{2 \cdot r}, v\right)$$

#7:
$$v = - e \cdot \sqrt{\left(\frac{k}{m \cdot r}\right)} \ \lor \ v = e \cdot \sqrt{\left(\frac{k}{m \cdot r}\right)}$$

#8:
$$v^2 = \frac{e^2 \cdot k}{m \cdot r}$$

#9:
$$\left(\frac{h \cdot n}{m \cdot r}\right)^2 = \frac{e^2 \cdot k}{m \cdot r}$$

#10:
$$SOLVE\left(\left(\frac{h \cdot n}{m \cdot r}\right)^2 = \frac{e^2 \cdot k}{m \cdot r}, r\right)$$

#11:
$$r = \pm\infty \ \lor \ r = \frac{h^2 \cdot n^2}{e^2 \cdot k \cdot m}$$

orbits are discrete=quantized. Orbit is determined by the n integer. n=
1,2,3...

Note r=∞ then E=0.
Minimum energy to remove the electron=ionization energy: H in ground state=
13.6 eV

Bohr radius=orbit with smallest radius, a(0).: n=1 a(0)=.0529 nm
h,e,k,m,n are all constants.

#12:
$$\frac{63}{1195417600000}$$

#13:
$$5.27012485 \cdot 10^{-11}$$

r(n)=n^2÷a(0) any orbit. (m)

#14: $r = n^2 \cdot a$

#15:
$$r = \frac{529 \cdot n^2}{10000000000000}$$

#16:
$$r = 5.29 \cdot 10^{-11} \cdot n^2$$

#17: $r = n^2$

The radius increases by the square of the integer. The Order of Magnitude is
in
10^(-9)m. Very small.

Sub r(n) into E:E(n) = –k(e)e^2/2a(0)n^2 n=1,2,3...
E(n) = – 13.606/n^2 eV   n=1,2,3... Energy is quantized.
Lowest energy state=ground state: n=1 E(1)= –13.606 eV –> +13.606 eV needed
as
ioniziation energy (energy into the system) to remove the electron.
                    excited state: n=2 E(2) = E(1)/2^2= –3.401 eV

#18: $e = - \dfrac{k \cdot e^2}{2 \cdot a \cdot n^2}$

#19:
$$- \frac{5766068187}{2635000000000000000000000000 \cdot n^2}$$

#20:
$$-\frac{2.188261171 \cdot 10^{-18}}{n^2}$$

#21:
$$-\frac{2.188261171 \cdot 10^{-18}}{n^2} \cdot \frac{1}{1.602 \cdot 10^{-19}}$$

#22:
$$-\frac{13.65955787}{n^2}$$

1 e = 1 eV

Bohr, by introducing the term:n*h, has quantized the H atom. This made
the particle nature of the atom more discrete, finite and determined,a
way of thinking of particles which is natural. The H atom is made of
subatomic particles:p+, n0, and e-. The p+ defines the element. The e-,
especially in valence shells, defines the reactivity of the atom.He
described the position, velocity, acceleration, and energy as being
quantized.By using classcial mechanics and treating the atom as a particle,
this explained some atomic physical properties of the atom. This science
has greatly added to the understanding towards the ideas of electron shells,
energy levels, and chemical reactivity.

The electron occupies an orbit based on an energy level which is discrete.

If the change in energy during a transition is positive, the electron
absorbs energy and moves farther way from the nucleus. And if large
enough, the energy change will reach ionization level. The atom will
eject the electron.
The opposite can occur: energy change is negative, electron emits energy,
and moves closer to the nucleus.
In the excited state (energy absorption), chemical reaction can occur.

On emission, all the photons create a spectral line.
All the lines create an emission sprectrum: all possible transitions.

The photon is a form of electromagnetic radiation: one important form is
infrared light (heat). λ is between 650-760 nm (one end of the visible
spectrum). In biological systems the radiation is absorbed in the
aquaeous environment of the cell and systems due to the excellent property
of water: specific heat.

$E(n) = -B/n^2$

$\Delta E(level) = E(f)-E(i)$ final-initial $n(f), n(i)$.

If $n(f)>n(i) \longrightarrow \Delta E$ is +.

If $n(f)<n(i) \longrightarrow \Delta E$ is -.

Planck's equation: $\Delta E(level)= h\nu$

#1: $e = h\cdot\nu$

#2:   SOLVE$(e = h\cdot\nu, \nu)$

#3: $$\nu = \frac{e}{h}$$

$c = \nu\lambda$

#4: $c = \nu\cdot\lambda$

#5:   SOLVE$(c = \nu\cdot\lambda, \lambda)$

#6: $$\lambda = \frac{c}{\nu}$$

#7: $$e = 2.179\cdot10^{-18}\cdot\left(\frac{1}{a^2} - \frac{1}{b^2}\right)$$

#8: $$e = -\frac{2179}{7200000000000000000000}$$

#9: $$e = -3.026388888\cdot10^{-19}$$

$\Delta E(level)$ between n=3 and n=2 which could apply to H, C, and other atoms.

#10: $$\nu = \frac{1513000000000000000}{3313}$$

#11: $v = 4.566857832 \cdot 10^{14}$

frequency of photon emission.

#12: $\lambda = \dfrac{3}{4566800}$

#13: $\lambda = 6.569151265 \cdot 10^{-7}$

wavelength of radiation=infrared spectra=heat 656 nm

Biochemical and organic reactions emit infrared radiation.

The Balmer series results in visible light including the above infrared light.

Duality of Matter: Particle and Wave

From the above discussion, Bohr studied the particle model of the H-atom. For instance, the momentum (p=mv) and Energy. However, this could not explain all the physical nature of the atom. For this problem, there was another model needed, the wave model (wavelength and frequency). De Broglie first introduced this model. By examining to simple equations, you can see this Duality character of the atom.

Momentum: (m*v=p) = Planck's constant (h) / wavelength (Lambda)

E (energy) = Planck's constant (h) * f (frequency)

Particle is proportional to wave by Planck's constant.

Position and Velocity is Uncertain

One final thought on the subject, Heisenberg's uncertainty principle is in effect when measuring the position or momentum of an electron.

Delta x * Delta momentum is greater than or equal to Planck's constant / (4 pi)

Planck's (bar) constant/2

Magnitude of 5.3 10^(-35) J*s

With an electron mass being 9.11*10^(-31)kg, this makes uncertainty large.

If the wavelength is known, then the momentum is known (DeBroglie). A single wavelength would exist throughout space. All points along the wave are the same. We have infinite uncertainty in the position of the particle. There is a range of positions.

If the momentum is a range, there is a range of wavelengths. The particle is a combination of wavelengths creating a wave packet. This range of positions is distinct than the rest of space. If momentum is lost all-together, we would add all wavelengths which would result in a wave packet of 0 length. We now have a distinct x position.

The uncertainties arise from the quantum structure of matter.

The Quantum model uses the wave model to further explain the H-atom.

The quantum particle can be an atom, proton, neutron or electron,
a particle of atomic or subatomic size or mass.

The particle will take the form of a wave.

The general math formula for such a wave is: y = Acos(kx-wt)

A=amplitude
θ=kx-wt
k=2π/λ
w=2πf
x=x coordinate on x axis
t=t, 2nd variable
y= 3rd variable on y axis

-w means wave is traveling in the + x axis direction

Graph is in the x-y plane

y depends on both variables x and t.

A particle is localized, found at single point.

A wave is unlocalized, found everywhere is space.

The combining of 2 waves leads to interference: constructive and
destructive.

A constructive region is called the wave packet which is considered
the particle. This is a distinct region of the wave form which repeats.

The wave evelope is the beat to this repetitive region of the wave form.

The derivation of the wave packet takes 2 waves with the same A and
different f. Add them by superposition principle. Use a trig identity
to find the wave packet. In the end, we have to cos functions multiplied
together. The first term is the beat, the envelope function.

y1

#1: $4 \cdot \cos(0.043 \cdot x - 25.3 \cdot t)$

t=1s

#2: $4 \cdot \cos\left(\dfrac{157 \cdot x}{1000} - \dfrac{503}{10} \right)$

y2

#3: $4 \cdot \cos(0.2 \cdot x - 75.6 \cdot t)$

t=1s

#4: $\quad 4\cdot\cos\left(\dfrac{x}{5} - \dfrac{378}{5}\right)$

y=y1+y2

#5: $\quad 4\cdot\cos\left(\dfrac{157\cdot x}{1000} - \dfrac{503}{10}\right) + 4\cdot\cos\left(\dfrac{x}{5} - \dfrac{378}{5}\right)$

the envelope function

#6: $\quad 8\cdot\cos\left(\dfrac{43\cdot x}{2000} - \dfrac{253}{20}\right)$

The y function with the wave packet

#7: $\quad \left(8\cdot\cos\left(\dfrac{43\cdot x}{2000} - \dfrac{253}{20}\right)\right)\cdot\cos\left(\dfrac{0.157 + 0.2}{2}\cdot x - \dfrac{50.3 + 75.6}{2}\cdot t\right)$

t=1s

#8: $\quad 8\cdot\cos\left(\dfrac{43\cdot x}{2000} - \dfrac{253}{20}\right)\cdot\cos\left(\dfrac{357\cdot x}{2000} - \dfrac{1259}{20}\right)$

You can calculate phase speed (single wave) and group speed (envelope speed).
   The evelope represents the particle. Using kinetic energy and momentum we can show the group speed of the wave packet is the speed of the particle.

Wave packet=envelope function=beat=particle!

The electron is a quantum particle with wave like properties.
The particle is moving to right with repeating position along the x axis.
You can not know with certainty the exact position and velocity of the particle at the same time. Once you fixed the position, you create a standing wave, which it is not; it is a traveling wave. Once you find it's velocity, you lose it's position. The y function is a double variable type. You can see one variable by making the other variable constant. But this destroys the function.

Notes:

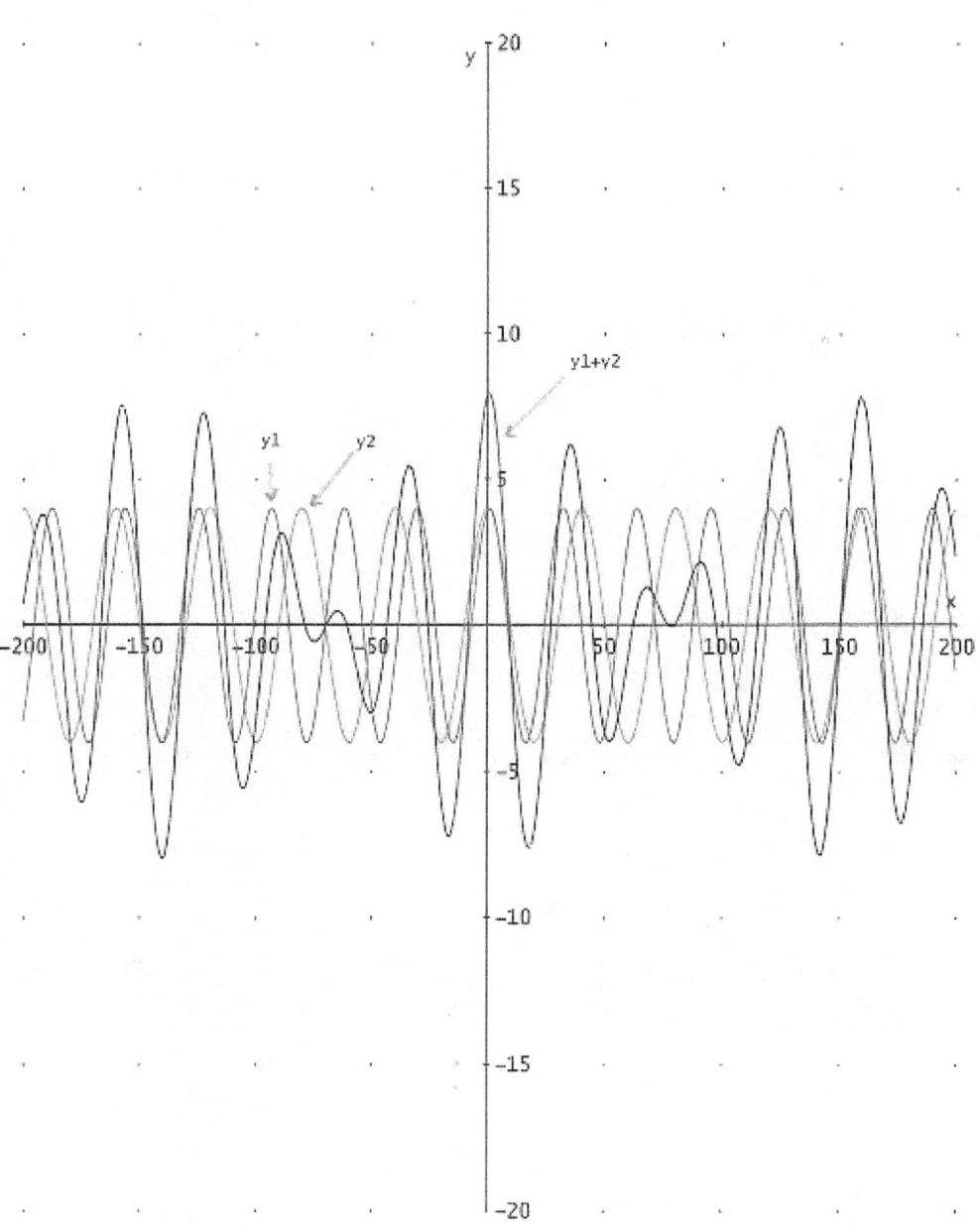

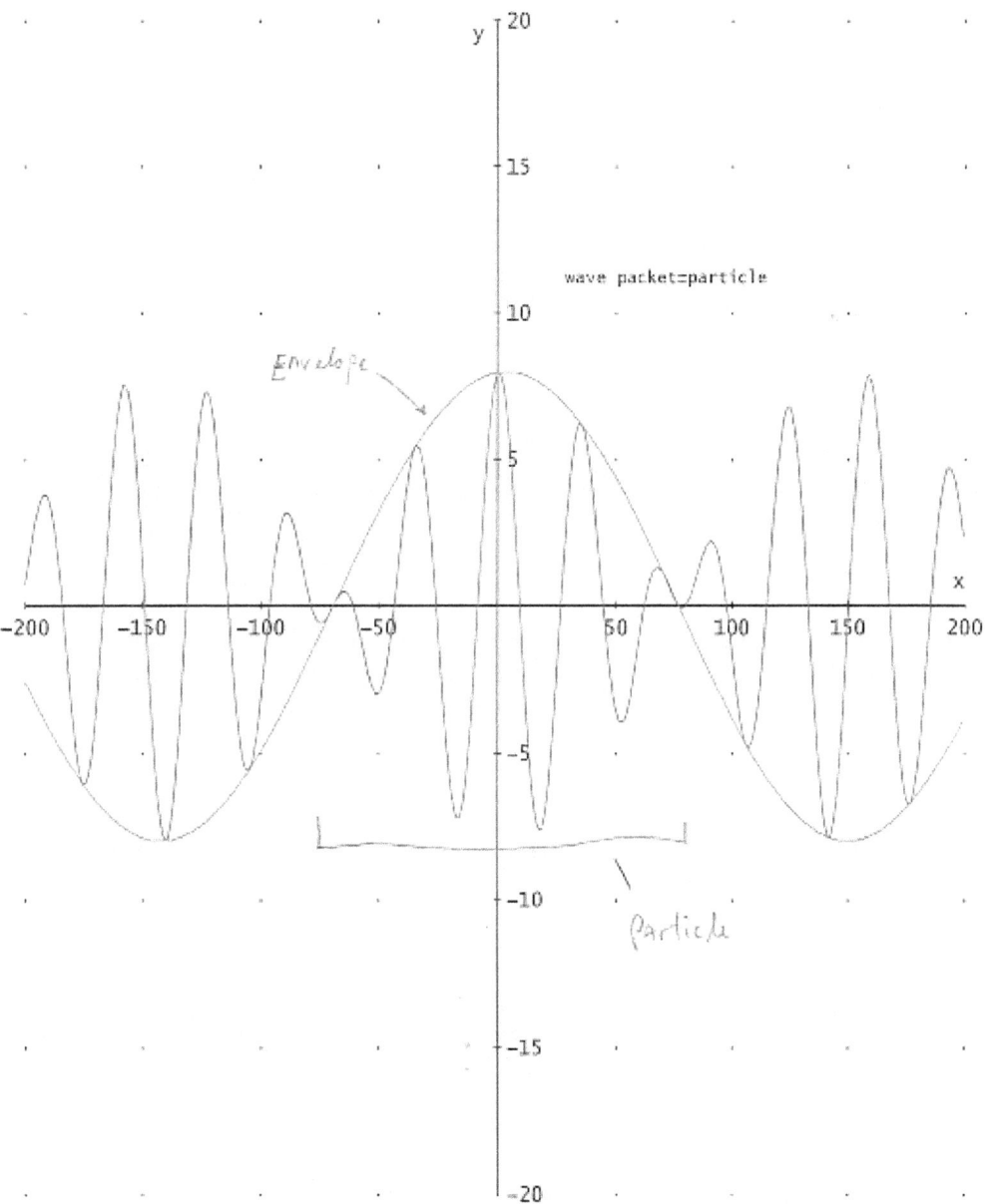

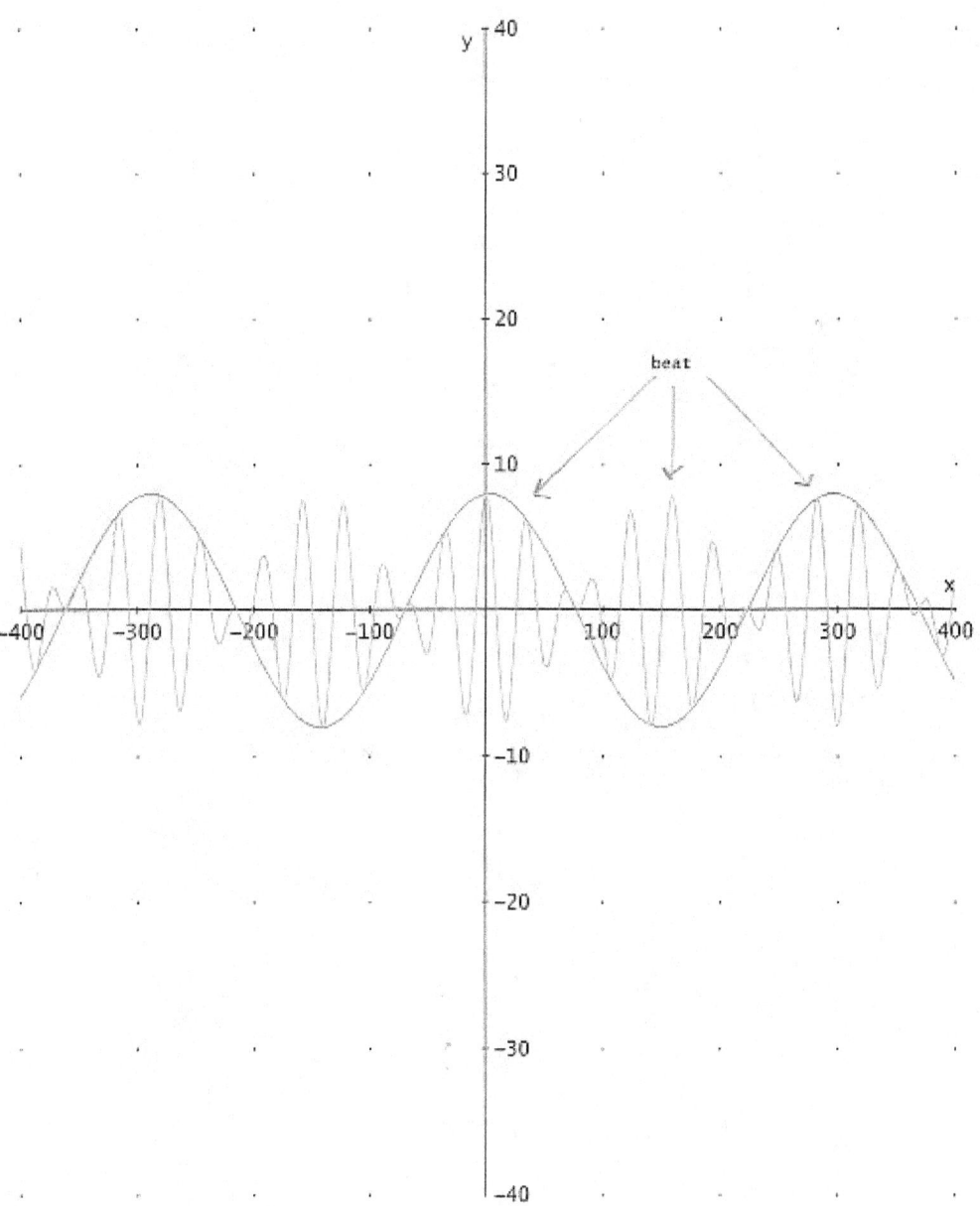

Let us try to locate an electron if we have it's velocity and accuracy of measurement. We can use the uncertainty principle.
speed of electron: 5*10^3 m/s to an accuracy of .003%:u
? minimum uncertainty in it's position.

This is a 1D motion along the x-axis.

#1:   p = m·v

momentum

#2: $p = \dfrac{911}{200000000000000000000000000000000}$

#3: $p = 4.555 \cdot 10^{-27}$

kg∘m/s

#4:   u·p

#5: $\dfrac{273}{200000000000000000000000000000000000}$

#6: $1.365 \cdot 10^{-31}$

u = uncertainty of p = kg*m/s

#7: $x \cdot u \geq \dfrac{h}{2}$

#8: $\text{SOLVE}\left(x \cdot u \geq \dfrac{h}{2},\ x \right)$

#9: $\text{IF}\left(u < 0,\ x \leq \dfrac{h}{2 \cdot u} \right) \vee \text{IF}\left(u > 0,\ x \geq \dfrac{h}{2 \cdot u} \right)$

#10: $x \geq \dfrac{21}{54800}$

#11: $x \geq 0.0003832116788$

m = .383 mm of position uncertainty
J∘s/(kg*m/s)=[kg*s*m^2/s^2]/(kg*m/s)
h = h/2π

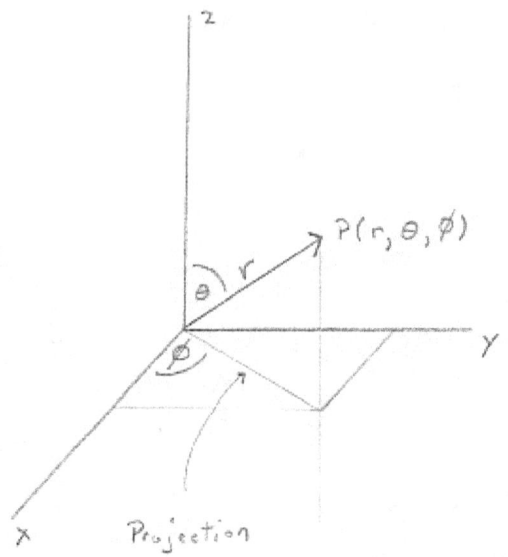

The principal quantum number, n, represents the allowed potential energy states of the H-atom.

The potential energy function for the H-atom:

U(r) = - k(e) e^2/r

E(n) = -(k(e) e^2/2a(0) * 1/n^2

    = - 13.606 eV/n^2 n = 1,2,3,...

n range: 1 –> (+) infinity = allowed values [0, (-) not allowed]

n is associated with the R(r), radial function: part of the full wave equation function.
R(r) will give functions that determine the probability of finding an electron at r distance from the nucleus.

1D wave equation: -h(bar)/2m * d^2 psi/dx^2 + U psi = E psi

h(bar) = h/2 pi

3D wave equation: for H-atom

-h(bar) /2m [(psi(x)" + psi(y)" + psi(z)"] + U psi = E psi

psi (x)" = 2nd derivative of psi.

rectangular coordinate: psi(x,y,z) conversion to polar: psi(r,theta, phi)

psi (r, theta, phi) = R(r) * f(theta) * g(phi)

Set boundaries on these 3 functions will lead to 3 different quantum numbers for each allowed state of the atom. These are restricted to integers. They correspond to 3 independent degrees of freedom: 3 space dimensions.

r = sq. rt of $x^2+y^2+z^2$

In the end, we have the sum of 3 separate probabilities describing the state of the H-atom: 1p+, 1n0, 1 e-. These are described by using wave mechanics.

The Wave Equation

Psi

What is Ψ?

Electron is a particle: It's speed is a fraction of light; whose particle is a photon.

Probability per given volume of space α N per volume.
N = number of particles

N/V α I
I = intensity = magnitude of concentration of particles per volume.
V = 3D space

Density is mass=particle per volume.
mass α N

I α E^2
E = electric field amplitude = magnitude.

With these proportionalities: P/V α E^2

P is the probability of finding a particle in a given volume.

Particle is associated with an electromagnetic wave (radiation).
de Broglie's wave-particle duality

P/V α A^2 (wave)
A^2 is a complex function: not measureable.
E^2 is measureable.

matter:particle:wave α E^2 α P

A^2 of wave associated with the particle is the probability amplitude = wave function = Ψ.

Ψ(r,t): r is the position vector.

space and time are separable variables. Both can be written as a product of space function:Ψ and complex time function.

$$\Psi(r,t) = \psi(r) \quad * \quad e^{(-i\omega t)}$$
          position   time
          space
ω = 2πf : angular frequency of wave function
$i = \sqrt{(-1)}$

$|\Psi|^2 = \Psi' * \Psi$
always real and +.
Ψ' = complex conjugate of Ψ.

Macroscopic particle = classic mechanics
Microscopic particle = quantum mechanics
Max Born developed the probabilistic interpretation.
Schrodinger developed the wave equation.

STD form complex number: a + ib   rectangular form
a, b are constants.
a is real
b is imaginary
$i = \sqrt{-1}$
Can be plotted on a complex plane. i v. x.

$e^{(i\theta)} = \cos \theta + i \sin \theta$ = to STD form.

#1: $e^{i\cdot\theta}$

#2: $\cos(\theta) + i\cdot\sin(\theta)$

#3: $a + i\cdot b$

#4: $(a + i\cdot b)\cdot(a - i\cdot b)$

     complex number   complex conjugate

#5: $a^2 + b^2$

real (2 constants) and positive value (squared).

#6: $\cos(-\omega\cdot t)$

#7: $\cos(t\cdot\omega)$

They are the same!

#8: $i\cdot\sin(-\omega\cdot t)$

#9: $-i\cdot\sin(t\cdot\omega)$

They are different!

#10: $\cos(\theta)$

#11: $\cos(-\theta)$

There is no change in sign of cos because of right triangle geometry.

#12: $\sin(\theta)$

#13: $\sin(-\theta)$

#14: $|\sin(\theta)|^2$

#15: $|\sin(-\theta)|^2$

If you plot these functions, you will see what may be beats, envelope

function,
and maybe a wave packet.
Where they are both zero, could be a node (P=0).
Where they are maximum, could be P=1.Note it is at the mean or 1/2 distance
along the x axis.
Plot the absolute value squared to see the density function. Use −θ and get
a + value.

de Broglie relates momentum to wavelength. Wave function for a free particle
is:$\Psi(x) = A e^{(ikx)}$.
$k = 2\pi/\lambda$ = angular wave number
A = magnitude

Full wave function is: $\Psi(x,t) = A e^{(-ikx)} * e^{(-iwt)} =$
$A[\cos(kx-wt) + i\sin(kx-wt)]$
real part = wave packet

Ψ can not be measured.
$|\Psi|^2$ = is measureable. = probability density: the probability of finding
the particle in a small volume of space.
dV = small volume surrounding a particle then P = density * dV.

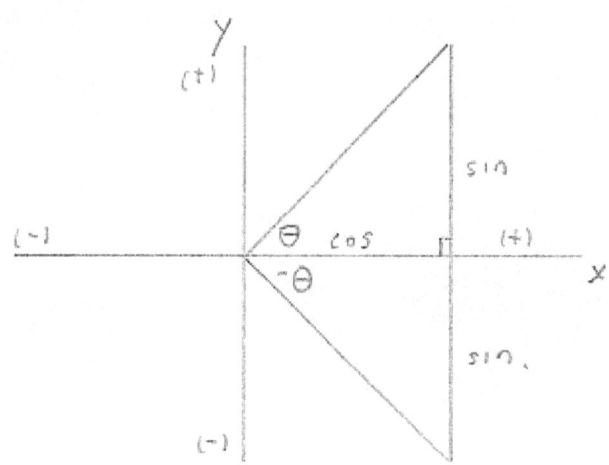

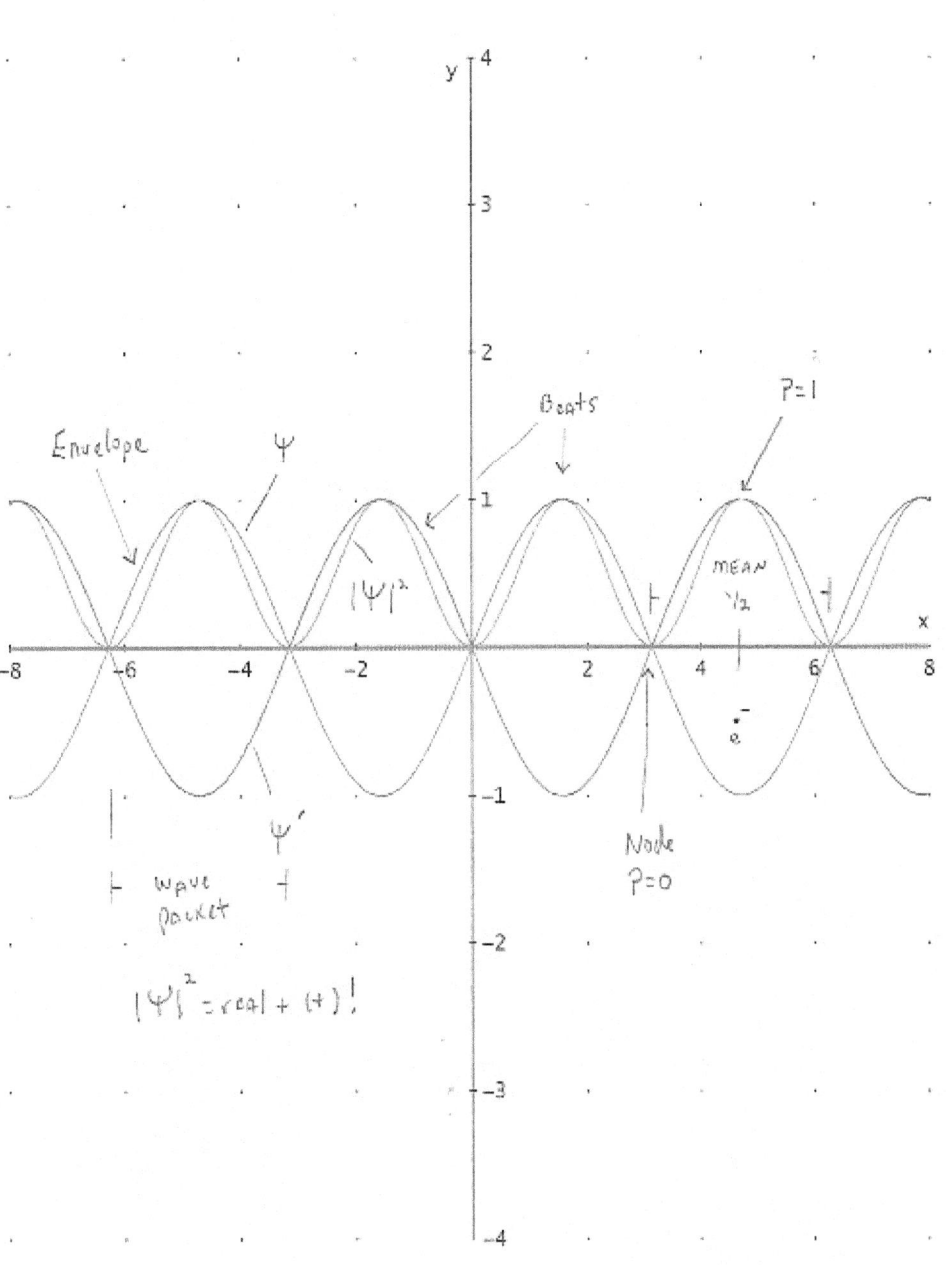

Envelope

Ψ

Beats

$P=1$

$|\Psi|^2$

MEAN
½

Node
$P=0$

Ψ'

wave
packet

$|\Psi|^2 = real + (+)\,!$

Expectation value

<x> : expectation value : weighted average position (location) of particle

Probability = |Ψ|^2 * dV : density in a given volume of space.

If 1D : P(x)*dx = |Ψ|^2*dx

If there is a given interval: a≤x≤b : P(ab) = ∫a->b |Ψ|^2 *dx = area under curve.

Σ of P along x axis : ∫-∞->∞ |Ψ|^2*dx=1 1=100% probability.
Satisfy (=1) means particle exists = Normalization.

Wave equation satisfied by Ψ is the Schrodinger equation.

Measureabe quantities of a particle: energy, momentum, weighted average position can be calculated.

<x> = ∫-∞->∞ (ψ^*)   * x   * Ψ   * dx
            complex conjugate   variable   wave equation   difference
in x
                                      f(x)=operator on Ψ

(Ψ^*)   *   (Ψ) = |Ψ|^2

#1:   - b·x²
    ψ = a·e

        1.26 4

particle is the electron
Ψ(x) = wave function of x
a=amplitude
b=constant

#2:   - 4·x²
    1.26·e

symmetrical wave function due to -x^2: "U" shape concave down.
e^x is the most increasing function in existence. Also can be represented as a trig function with real and imaginary terms. We are interested in the real part only.

We will use an example to calculate amplitude and expectation value: position of the particle at x value to highlight graphically the interesting points.

To find the amplitude, the wave function has to be normalized: =1.

∫-∞->∞ |Ψ|^2 *dx = 1

The probability density is equal to 1.
The area under the curve is the probability.

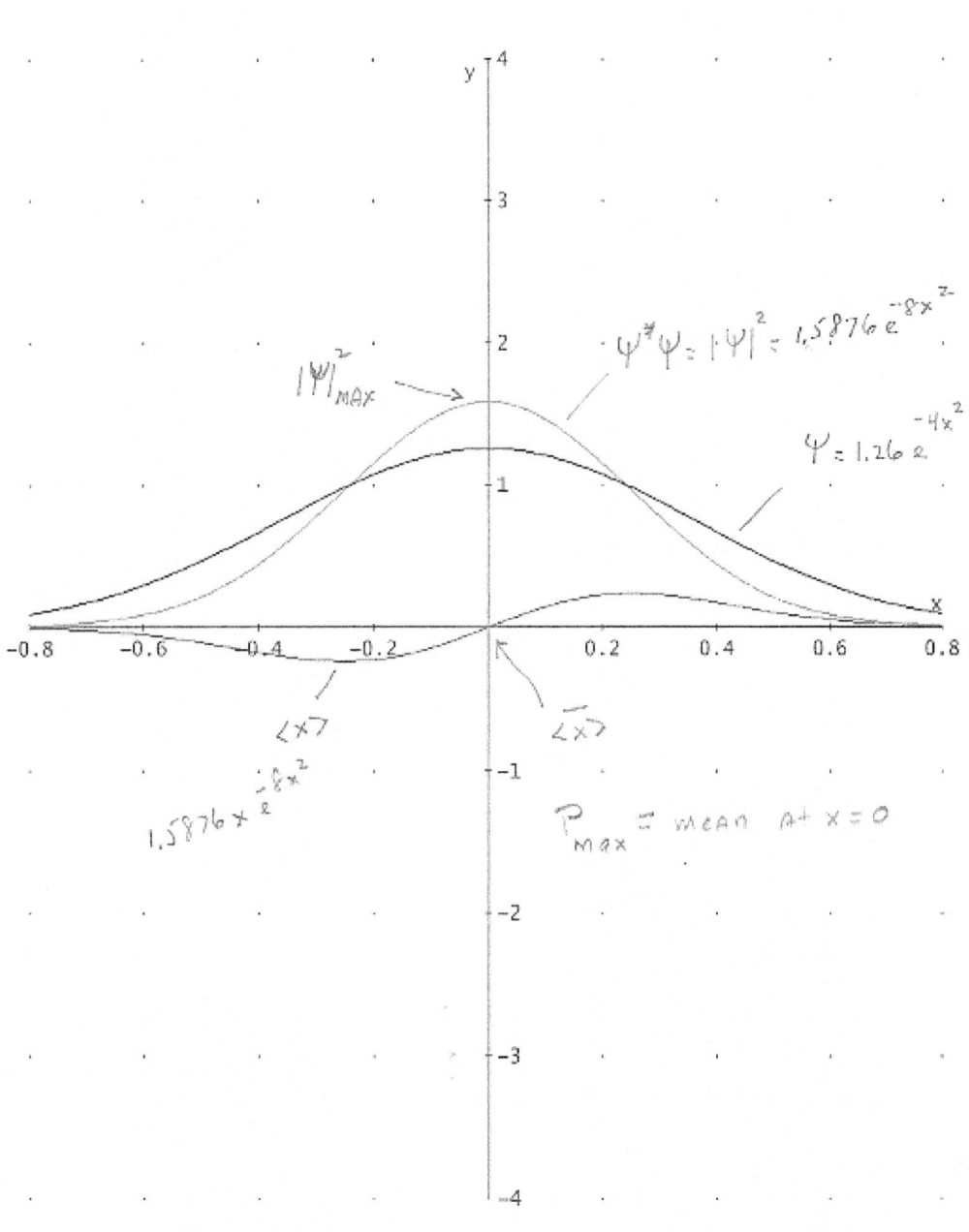

$|\Psi|^2_{MAX}$

$\Psi^* \Psi = |\Psi|^2 = 1.5876 \, e^{-8x^2}$

$\Psi = 1.26 \, e^{-4x^2}$

$\langle x \rangle$

$\langle x \rangle$

$1.5876 \, x \, e^{-8x^2}$

$P_{max} = mean \ at \ x = 0$

Probability Density

#3: $\left(a \cdot e^{-b \cdot x^2}\right)^2$

#4: $a^2 \cdot e^{-2 \cdot b \cdot x^2}$

#5: $1.5876 \cdot e^{-8 \cdot x^2}$

The probability density function.
The amplitude is squared.
The slope of x^2 is doubled.
The interval of the integral can be divided in 2: $-\infty \to 0 + 0 \to \infty$. This creates a sum.
The x^2 is (+) whether x is (+) or (-).
This creates a multiple of 2 * 1 interval ∫.
∫ is evaluated by Gauss's Probability Integral: $I(0) = \int 0 \to \infty\ e^{\wedge}(-ax^{\wedge}2) * dx = 1/2\ \sqrt{\pi/a}$

#6: $2 \cdot a^2 \cdot \left(\dfrac{1}{2} \cdot \sqrt{\left(\dfrac{\pi}{2 \cdot b}\right)}\right) = 1$

#7: $\dfrac{\sqrt{2} \cdot \sqrt{\pi} \cdot a^2}{4} = 1$

#8: $\text{SOLVE}\left(\dfrac{\sqrt{2} \cdot \sqrt{\pi} \cdot a^2}{4} = 1,\ a\right)$

#9: $a = -\dfrac{2^{3/4}}{\pi^{1/4}} \lor a = \dfrac{2^{3/4}}{\pi^{1/4}}$

#10: $a = -1.263237555 \lor a = 1.263237555$

#11: $\psi = 1.26 \cdot e^{-4 \cdot x^2}$

$<x> = \int -\infty - \infty\ (\psi^{\wedge}{*}) * x * \psi * dx$
$\psi^{\wedge}{*}$ = conjugate of ψ
$\psi = \psi$
Both multiplied together gives a real and positive value: $\Psi^{\wedge}2$
x = x (variable of 1D)
dx = change in x.

Again the amplitude is squared.
The slope of x^2 is doubled.

#12: $a^2 \cdot x \cdot e^{-2 \cdot b \cdot x^2}$

The interval is divided in 2 creating a sum.
For the second $\int$ change the x -> -x: -∞->0 --> ∞->0.
Reverse order of limits creates a (-) sign.
With this, the sum is 0 = <x>: at x=0, average position is at x=0.

Note that the expectation value is vertically aligned with the probability
density maximun. This is the best place of finding the particle on the x
axis.

#13: $1.5876 \cdot x \cdot e^{-8 \cdot x^2}$

Notes:

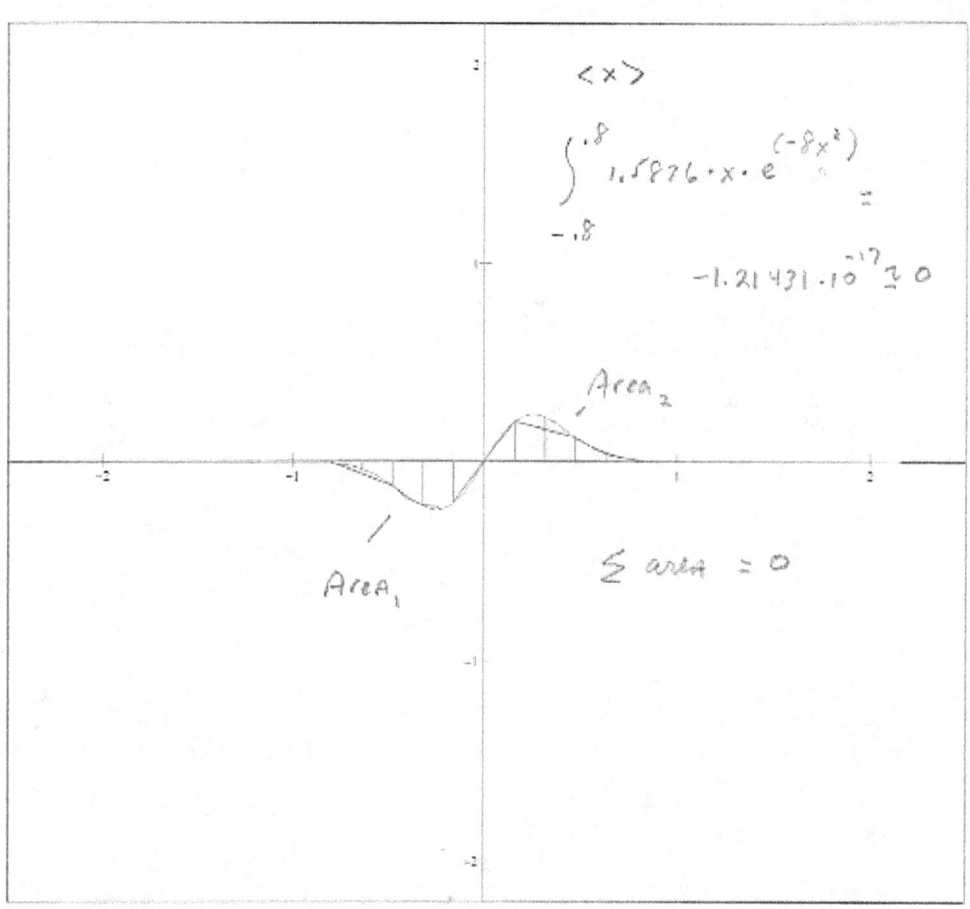

$$\langle x \rangle$$

$$\int_{-.8}^{.8} 1.5876 \cdot x \cdot e^{(-8x^2)} =$$

$$-1.21431 \cdot 10^{-17} \doteq 0$$

Area$_2$

Area$_1$

Σ area $= 0$

Notes:

#1: $\dfrac{d}{dx}\left(1.5876 \cdot e^{-8 \cdot x^2}\right)$

#2: $-\dfrac{15876 \cdot x \cdot e^{-8 \cdot x^2}}{625}$

Derivative of $|\Psi|^2$

#3: $-25.4016 \cdot x \cdot e^{-8 \cdot x^2}$

#4: $1.26^2 \cdot x \cdot e^{-2.4 \cdot x^2}$

#5: $1.5876 \cdot x \cdot e^{-8 \cdot x^2}$

<x> value: Similiar to $(|\Psi|^2)'$

Notes:

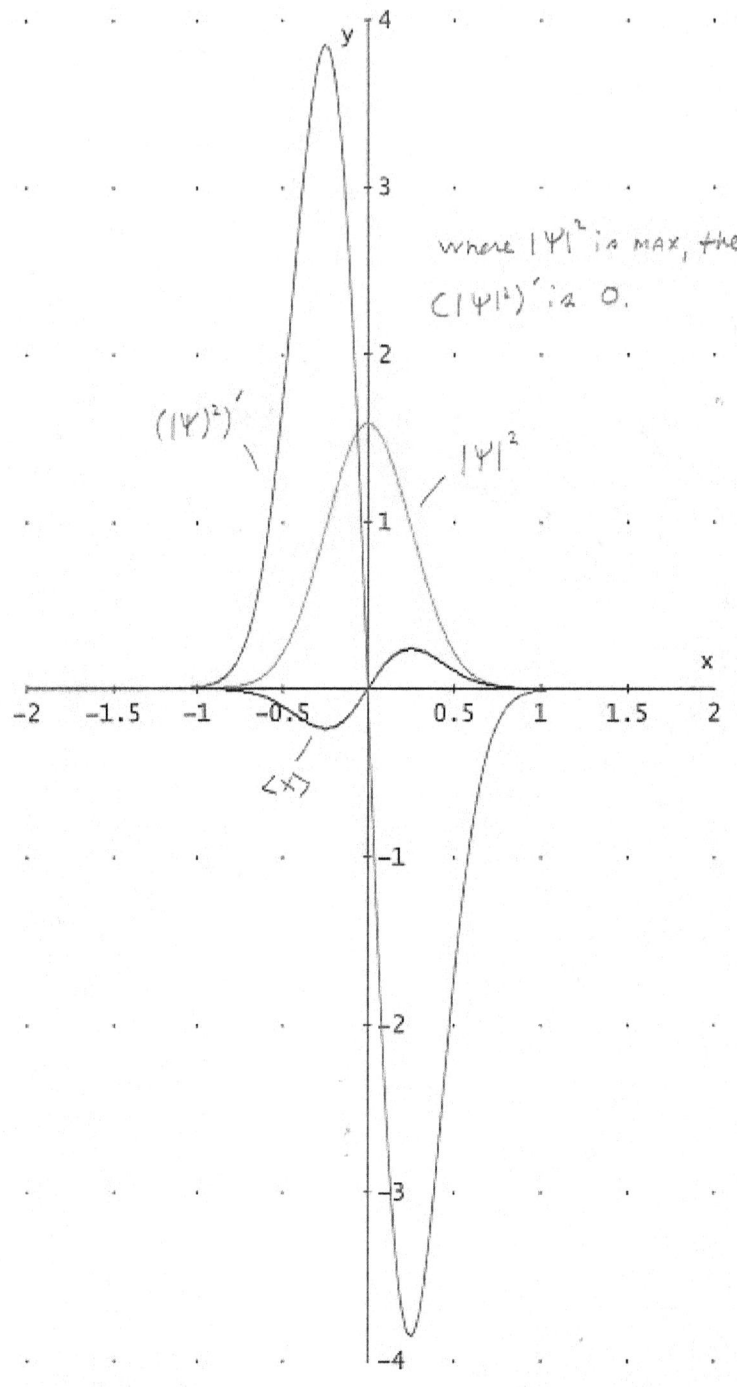

where $|\Psi|^2$ is max, the $(|\Psi|^2)'$ is 0.

$(|\Psi|^2)'$

$|\Psi|^2$

$\langle x \rangle$

1 Principal n

Energy-level diagram: Model is the particle in a box (1D).

The box is an energy one with U=0 inside the box. The Total energy in the box is all Kinetic.

x axis length (L) is π.

The walls of the box have U=∞ and the Ψ=0 (boundaries).

The energy of the particle is quantized: an integer of n.
The ground state (the lowest) is 1.
n=2,3,4.. are the excited states.
E=0 does not exist. The particle can never rest.

Ψ(x)=A∗sin(n∗x) : wave equation

|Ψ|^2 : probability density

Using momentum and kinetic energy arguments, the allowed energy states are derived.
Energy levels are discrete, allowed values, meaning quantized.

E increases with increasing n.

#1: SIN(x)

Ψ1 = n1 = E1 = 1^2E1

#2: SIN(2·x)

E2=4E1 = 2^2 E1

#3: SIN(3·x)

E3=9E1 = 3^2E1

Ψ wave equation

#4: $|SIN(x)|^2$

#5: $|SIN(2·x)|^2$

#6: $|SIN(3·x)|^2$

|Ψ|^2 probability density

A bound electron means that the electron is held within 2 potential energy barriers. The barrier is impenetrable due to it's ∞ state at the walls (boundaries). If you know the L (distance=nm) between the 2 barriers, you can calculate the n states of energy levels. The electron takes on excited states within the walls of the potential energy barriers. As a wave function, this means the frequency increases and the period decreases.

#7: $\quad e = \dfrac{h^2}{8 \cdot m \cdot l^2} \cdot n^2$

$n=1,2,3,\ldots$ Energy α n^2

If L=.2nm ?E1 etc.

h=6.63*10^(-34)J*s Planck's constant
m(e)=9.11*10^(-31)kg

#8: $\qquad e = \dfrac{439569}{291520000000000000000000}$

#9: $\qquad e = 1.507851948 \cdot 10^{-18}$

Joules

#10: $\dfrac{1.507851948 \cdot 10^{-18}}{1.6 \cdot 10^{-19}}$

#11: $\qquad\qquad$ 9.424074675

eV

n=2 2^2=4

#12: 4·9.424074675

#13: $\qquad\qquad$ 37.69629869

eV

n=3 3^3=9

#14: 9·9.424074675

#15: $\qquad\qquad$ 84.81667207

eV

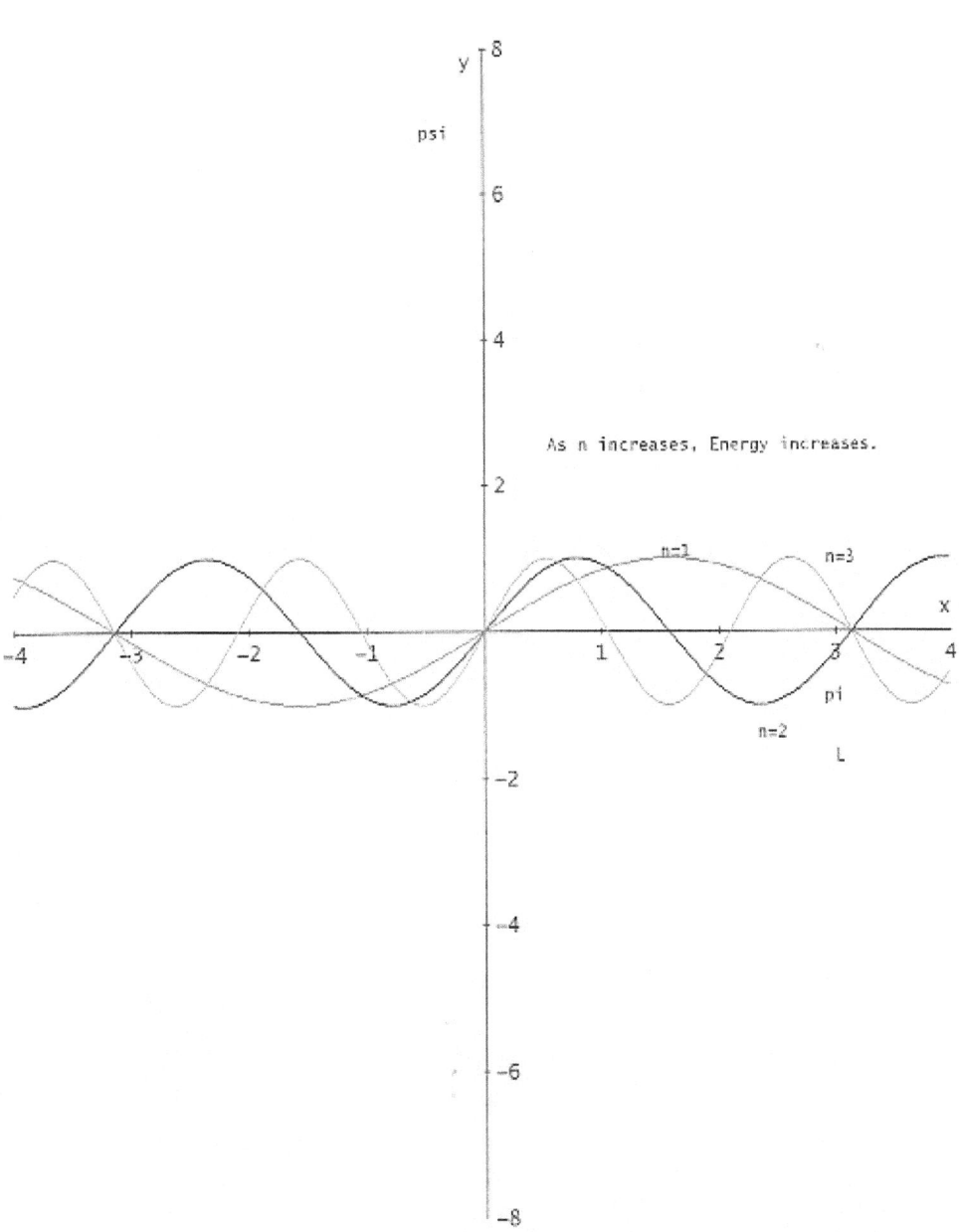

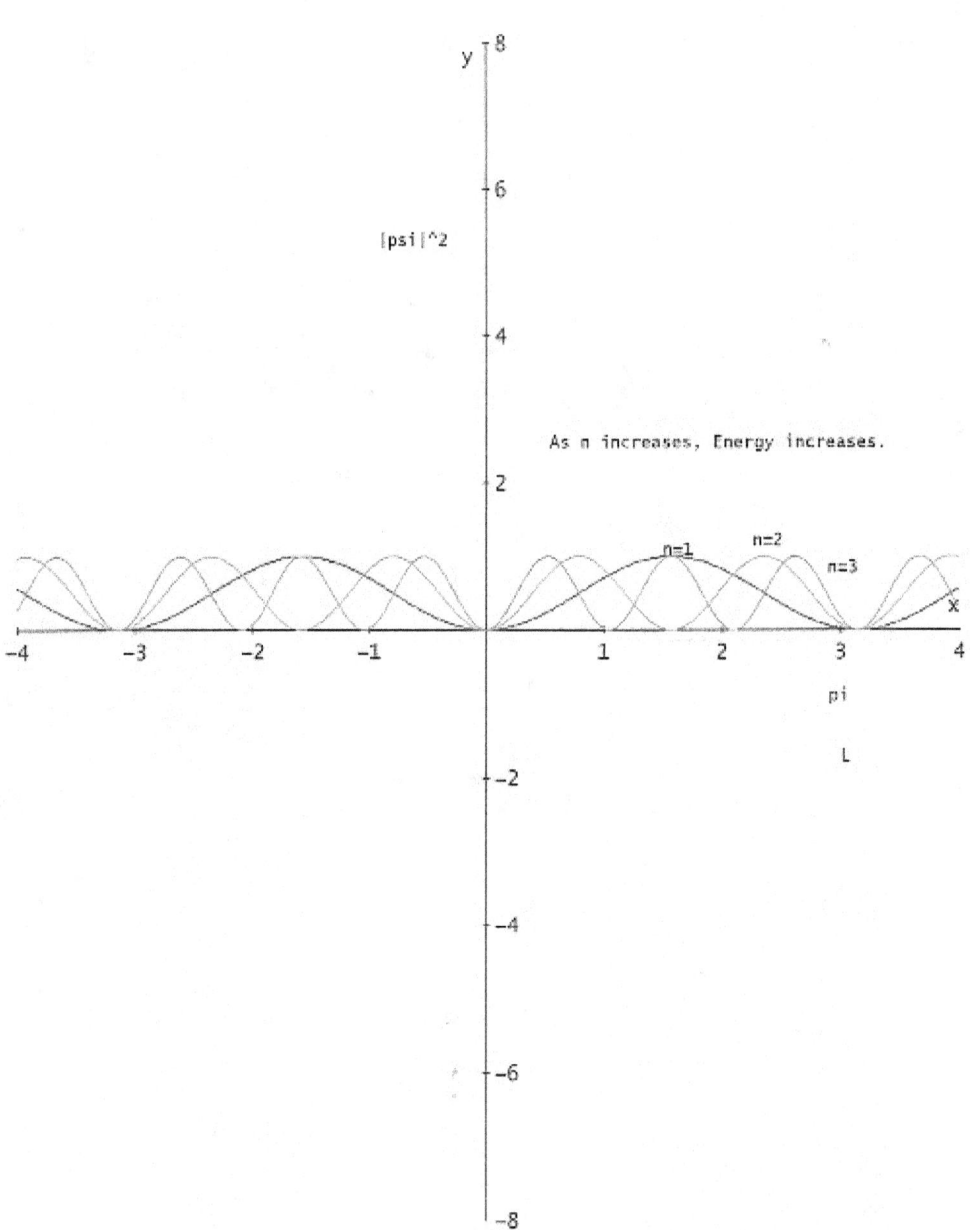

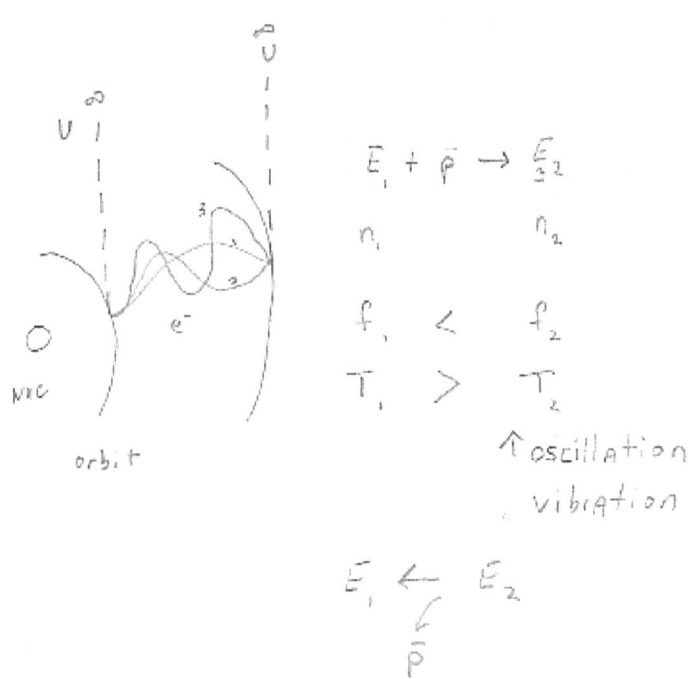

The model of the atom: Nucleus has protons p+ and neutrons n0. Electrons
e- orbit around nucleus. The force is electrical attraction. The inner
core electrons diminish the force. If r is the radius of atom or the
distance between proton and electron, the potential energy is inverse
squared to r : -(1/r^2). As r increases, the U decreases. If U is low
enough, the electron can escape (chemical reaction). Atoms absorb energy
and lower the U barrier. The energy absorbed can be radiation:light
(photon), heat (infrared light), chemical (ATP) or physical (twisting,
stretching).

To calculate the energy needed to excite the electron from state to
another (transition) would be to use ΔE=En-En-1.

You can caluculate the λ of the exciting radiation by using
λ=(hc)eV*nm/ΔE eV. So Planck's constant times light speed divided by
energy difference equals wavelength of exciting radiation (photon).

#1: $\lambda = \dfrac{h \cdot c}{e}$

hc/eV -> eV*nm   eV*nm/eV -> nm

   As mentioned above, n is associated with the radial function, R(r), of the full wave function.
R(r) → functions → probability of finding the electron at a certain radial distance from the nucleus.
Again, the potential-energy function, U(r), depends only on the r, radial coordinate. The energies of the

allowed states for H-atom are found with the function, $E(n) = -13.606/n^2$ eV n=1,2,3,...
There are 3 additional quantum numbers:

little L = orbital quantum number = orbital angular momentum
range : 0 → n-1

m subscript little L = orbital magnetic quantum number
range: - little L → little L

These describe the electron.
Both are integers.

eg. n = 2 → little L = 0 or 1 if little L = 0 → ml = 0, if little L = 1 → ml = -1,0,1.

m subscript little s = magnetic spin quantum number
range: ½ → - ½

All 4 numbers describe the allowed state of the atom.

eg.   n = 2
         l = 0 or 1
         ml = -1,0,1
         ms = ½ or - ½
              1e   1e   2 e in 1 orbit

total = 4 different allowed states of the atom: 1 2s + 3 2p
         2,0,0   2s
         2,1,-1   2p x
         2,1,0   2p y
         2,1,1   2p z

         2s = n=2, l=0,ml=0   n=2
         2p =   2   1   -1   l=n-1 :2-1=1 = 0,1
                2   1   0
                2   1   1   ml = 2l+1 = 2*1+1 =3 = -1,0,1

         All have the same energy = -3.401 eV

         All states with the same n = shell.

         n =1 = K shell
            2   L
            3   M
            4   N
            5   O

6   P

....

All states with the same n and l = subshell.

l = 0 = s (sharp)
    1 = p (principal)
    2 = d (diffuse)
    3 = f (fundamental)
    4 = g
    5 = h
....

Formulas for the allowed states of the atom:

Allowed values	# of Allowed states
n : 1,2,3,... infinity (1 → + infinity)	+ infinity
l : 0,1,2,..., (n-1) (0 → + infinity)	n
ml ; -l, -l+1,...,0,...l-1,l (-l,0,l)	2l+1

eg. n=1   l=0   ml=0
                1 state   1 state

n=2   l=0,1   ml= -1,0,1
          2 states   3 states

eg state 3p = n=3 + l=1

2s = n=2 + l=0

States that violate the rules do not exist. The boundary conditions on the wave function are not satisfied. Using n and l only, for n=2, there are 2s and 2p states allowed, not 2d or 2f. For n=3, 3s, 3p, 3d are allowed. In organic chemistry, we are concern about the n=1,2,3 states.

Upsi, potential-energy of the H-atom, depends only on the r, radial distance. f(theta) and g(phi) are constants.

$$psi(r,theta,phi) = R(r)* f(theta)* g(phi)$$
        c c   c   c   r is 1 variable

The simplest psi function: psi 1s(r) = 1s state = ground state (lowest energy state).

$$psi 1s (r) = [1/sqr.(pi a(0)^3)] * e^{[-r/a(0)]}   a(0)=Bohr radius = .0529 nm$$

Psi $\rightarrow$ 0 as r $\rightarrow$ infinity (electron escapes orbit creating an empty shell).

Psi is spherically symmetric (all s states have the same shape, just larger).

Psi needs to be normalized.

The 2 functions: probability density and function created.

The geometry is a 3D position vector. The 3 quantum numbers give the state of the atom.

The terminal point of the vector are 3: x,y,z. These can represent any 3 values or entities we
want.
                        n,l,ml

                        r,theta,phi

                        R(r), P(theta),F(phi) : multiple term functions.

The main function of the psi is the exponential function.

The single variable, r, creates the sphere geometry.

1s, 2s, and 3s, are enlarging spheres.

The density is wavy with peaks, local maximums. The largest maximum is the most probable
value of r, the distance from the nucleus. All a multiple of a(0): 1s=a(0), 2s=5a(0), and 3s=13a(0).

The R(r) is a decreasing exponential function.

The density is the same type with a steeper slope.

In polar coordinates, they are circles.

The density function is a inverted parabola along the axis of r. 1s = 1 parabola, 2s=2 parabolas,
and 3s=3 parabolas. All have points on r where the probability is equal to 0. These are nodes.

The most probable value of r and the mean value of r (expectation value) are not the same due
to the asymmetry of the probability density function. They can differ as much as over 50%, not very
good. 1 out of 2 chance of knowing the electron position, a 50:50. The fundamental reason is the
quantum particle. The traveling wave, a wave with a velocity and no position, is not the same as a
standing wave, a wave with position and no velocity. At the atom level, the r is too small and the
speed is too fast to comprehend.

Density is the mass per volume. With psi, it is mass of the radiation type per volume of space.

The density function is a combination function made from the probability density and the shell
volume of a sphere.

Notes:

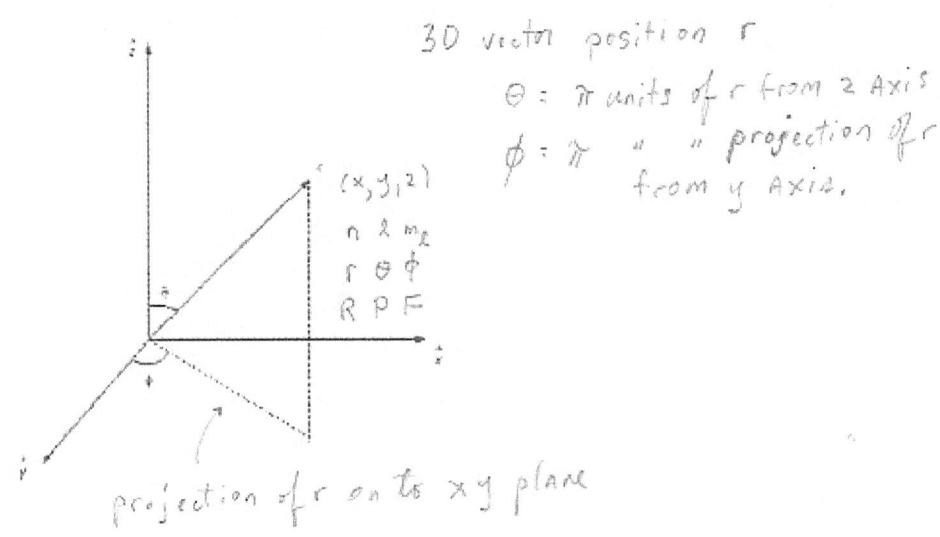

3D vector position r

θ = π units of r from z Axis

ϕ = π " " projection of r from y Axis.

projection of r on to xy plane

Notes:

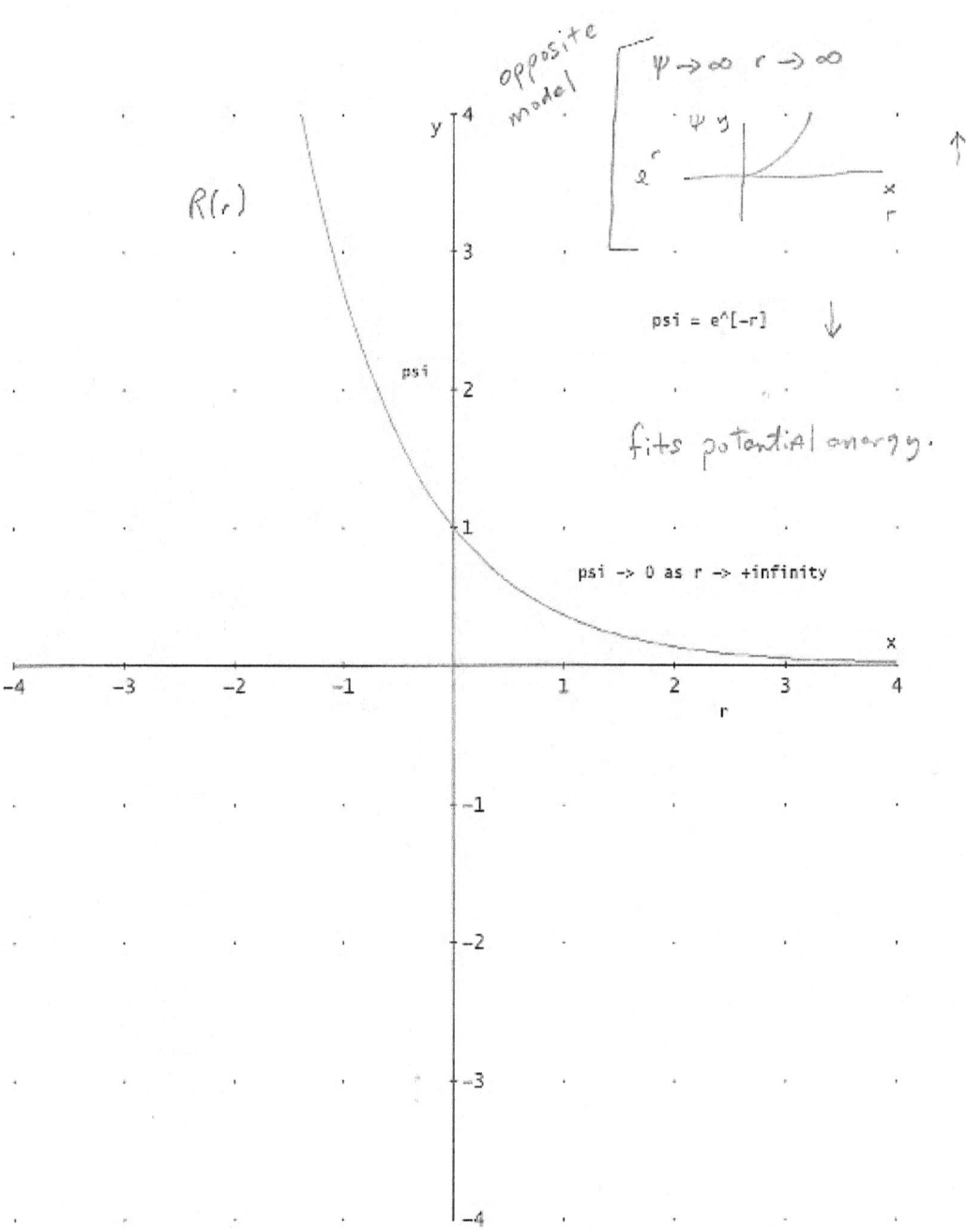

R(r)

opposite model

$\psi \to \infty$ $r \to \infty$

psi

psi = e^[-r]

fits potential energy.

psi -> 0 as r -> +infinity

Polar COORDINATES

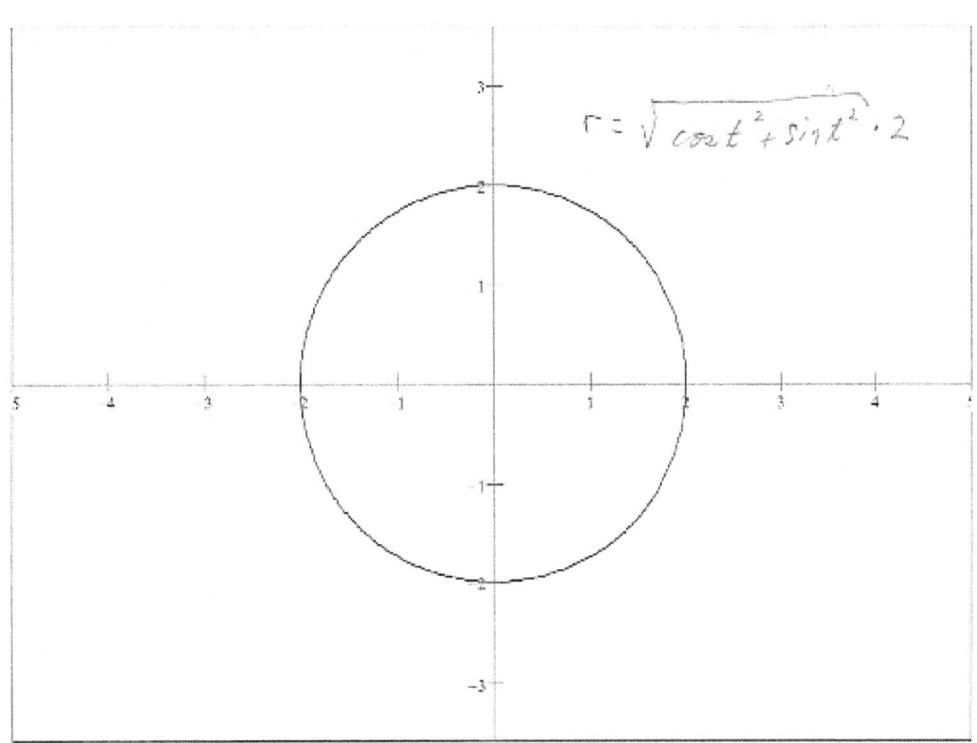

$$r = \sqrt{\cos t^2 + \sin t^2} \cdot 2$$

Notes:

```
ψ (n,l,ml)
   r θ φ
   1 0 0    1s
                  ms = +/- 1/2   2 electrons
```

$$\psi = F(\phi) \quad *P(\theta) \quad *R(r)$$

```
ψ = F(φ)      *P(θ)       *R(r)
   1/√2π      1/√2        2/a(0)^3/2 * e^[-r/a(0)]
    c          c          e^[-r]
```

#1: $\psi = \dfrac{1}{\sqrt{(\pi \cdot (0.0529 \cdot 10^{-9})^3)}} \cdot e^{-r/(0.0529 \cdot 10^{-9})}$

#2: $\psi = 1.466363208 \cdot 10^{15} \cdot e^{-1.890359168 \cdot 10^{10} \cdot r}$

Order of Magnitude of 10^15.

#3: $\psi = e^{-r}$

#4: $\dfrac{\sqrt{(2 \cdot \pi)}}{2 \cdot \pi}$

#5:   0.3989422804

#6: $\dfrac{0.3989}{\pi} \cdot 180$

#7:   22.85528644

φ = 22.86°

#8: $\dfrac{\sqrt{2}}{2}$

#9:   0.7071067811

#10: $\dfrac{0.707}{\pi} \cdot 180$

#11:   40.50811611

θ = 40.51°

#12: $\quad \psi = -2 \cdot e^{-r^2}$

#13: $\quad \psi = e^{-r^2}$

#14: $\quad \psi = -e^{-r^2}$

#15: $\quad \psi = 2 \cdot e^{-r^2}$

a(0) = h _ ^2/[k(e)*m(e)*e^2]
h_ = 1.054*10^(-34) J*s
m(e) = 9.109*10^(-31) kg
k(e) = 8.987*10^9 N*m^2/C^2
e = 1.602*10^(-19) C

#16: $\quad a = \dfrac{h^2}{k \cdot m \cdot e^2}$

#17: $\quad a = \dfrac{277729}{5252311511538300}$

#18: $\quad a = 5.287748058 \cdot 10^{-11}$

a(0) = .0529*10^(-9) nm 1st Bohr radius

Normalization: F*P*R

#19: $\quad \dfrac{1}{\sqrt{(2 \cdot \pi)}} \cdot \dfrac{1}{\sqrt{2}} \cdot \dfrac{2}{a^{3/2}} \cdot e^{-r/a}$

#20: $\quad \dfrac{e^{-r/a}}{\sqrt{\pi} \cdot a^{3/2}}$

ψ normalized.

#21:
$$\frac{100000000000000000000 \cdot \sqrt{10} \cdot e^{-10000000000000 \cdot r/529}}{12167 \cdot \sqrt{\pi}}$$

#22:
$$1.466363208 \cdot 10^{15} \cdot e^{-1.890359168 \cdot 10^{10} \cdot r}$$

Notes:

Probability density for the 1s state: |Normalized ψ|^2

#1: $\psi = \dfrac{1}{\sqrt{(\pi \cdot a^3)}} \cdot e^{-r/a}$

#2: $\left| \dfrac{1}{\sqrt{(\pi \cdot a^3)}} \cdot e^{-r/a} \right|^2$

#3: $\dfrac{0.3183098861 \cdot e^{-2 \cdot r/a}}{|a|^3}$

#4: e^{-r}

ψ 1s (r)

#5: $e^{-2 \cdot r}$

Normalized ψ 1s (r)

#6: $4 \cdot r^2 \cdot e^{-2 \cdot r}$

P 1s (r) spherical shell of radius r and thickness dr.
dV   = 4πr^2   *dr
Volume   Surface area   *thickness

Ave. value of r = expectation value of r = 3/2 a(0) = 0.810 nm.

<x> > 50% of a(0).
.81 nm   .53 nm
.81-.53 = .28 = Δ
.53/2 = .265   50% of a(0).
.28 > .265 nm
Asymmetry of R(r) : more area to the right of a(0).

Quantum mechanics says there is no sharp boundary. q, charge of the
electron, of the electron is extended throughout a diffuse region of
space = electron cloud.

The R(r) plotted as a function of r shows the darkest portion of
R(r) appears at a(0) = most probable value of r for the electron.

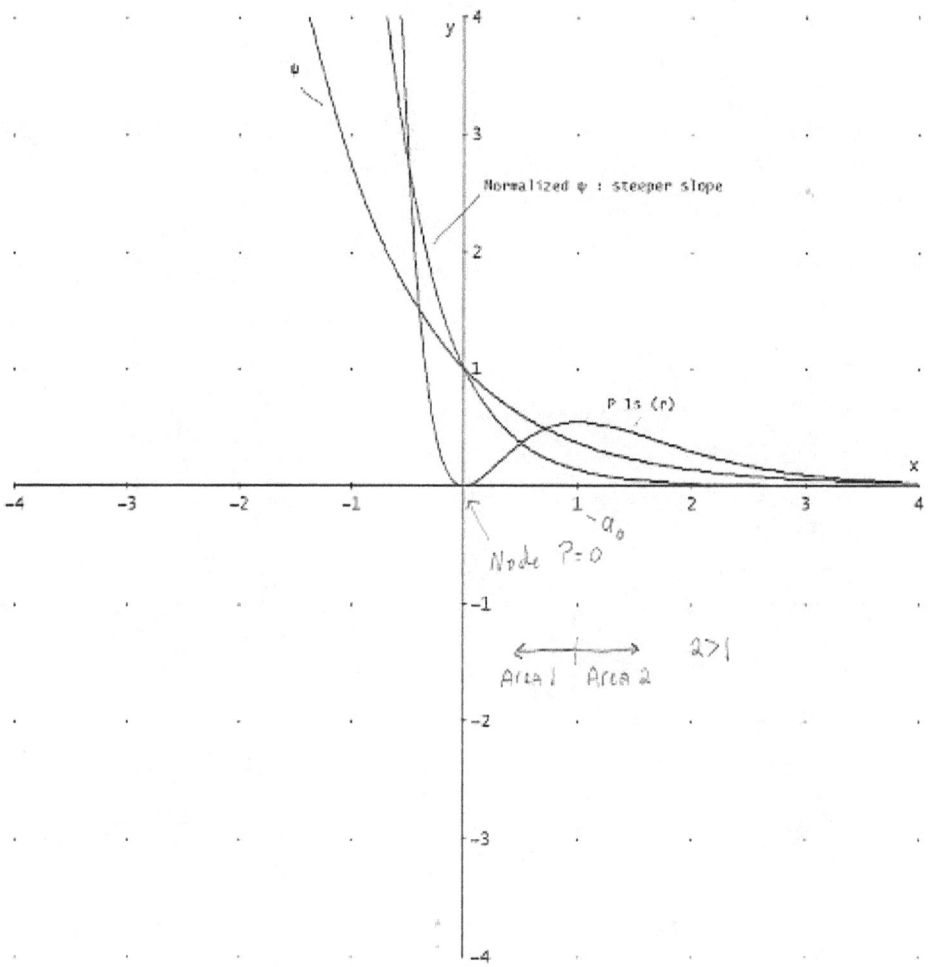

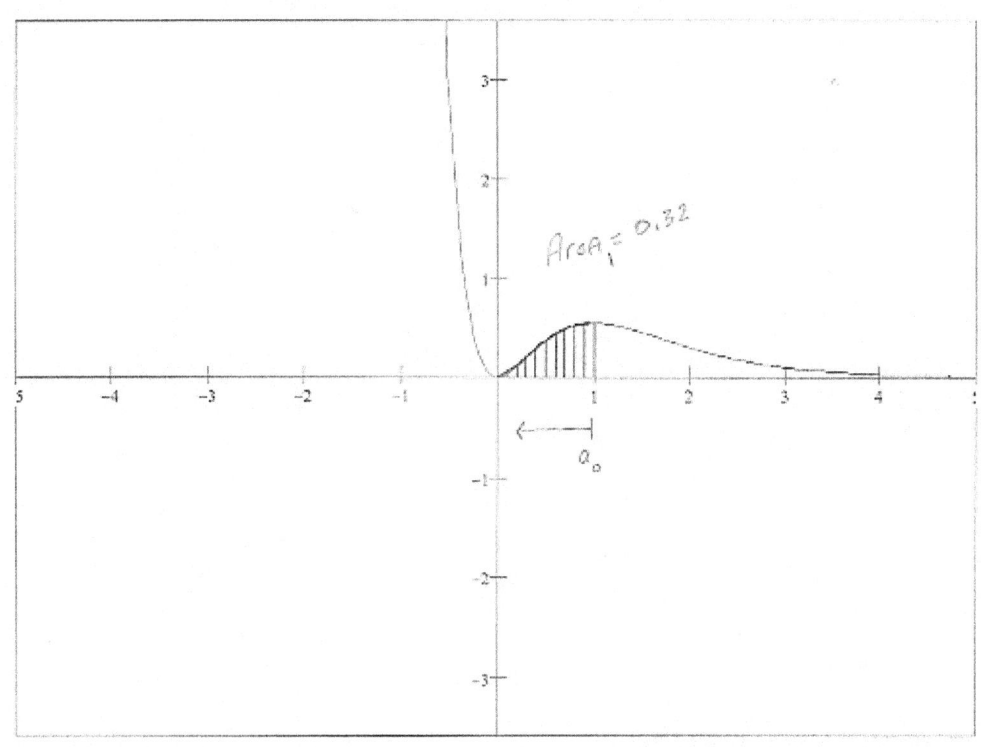

$Area_1 = 0.32$

a_0

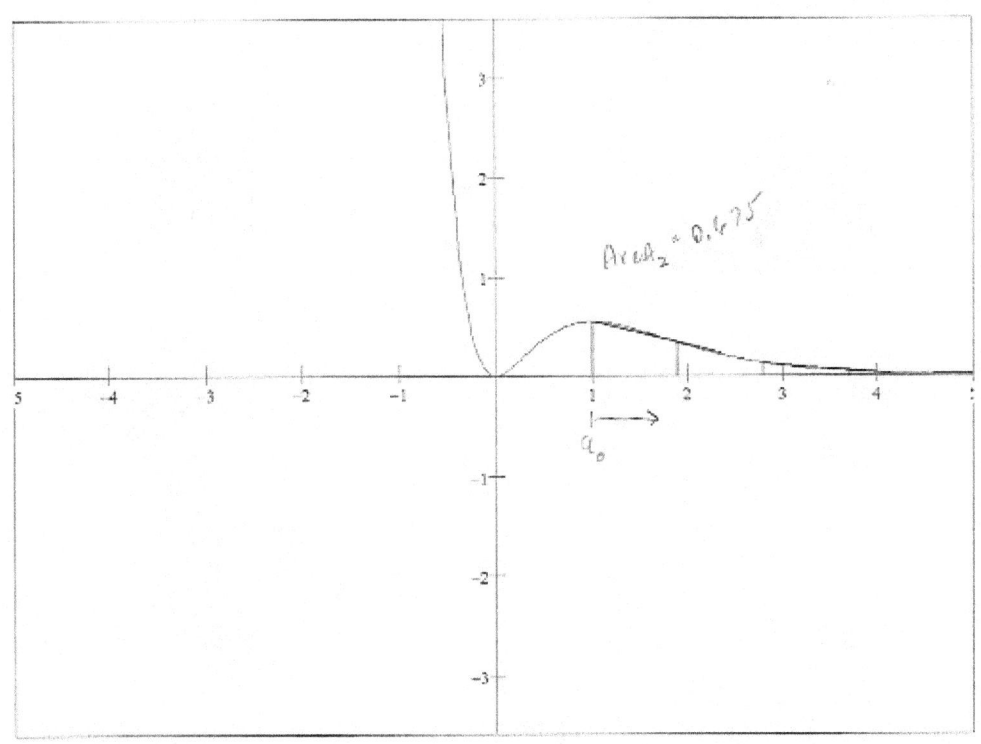

$Area_2 = 0.675$

a_0

Hydrogen Wave function Simulation

This simulation calculates the wave functions for hydrogenic (hydrogen like) atoms for quantum numbers n = 1 to n = 50. The upper left window shows the angular wave function, the upper right window shows the radial wave function and the lower left window shows a plot of the probability density (wave function squared) in the x - z plane.

Much like the case of an electron trapped in a one dimensional well the electrons of hydrogen like atoms are trapped around the atom in the coulomb potential. The wave functions representing the electron therefore have to 'fit' in the coulomb potential with the right boundary conditions.

You can enter various numbers of the three principle quantum numbers n, l and m in the window on the lower right.

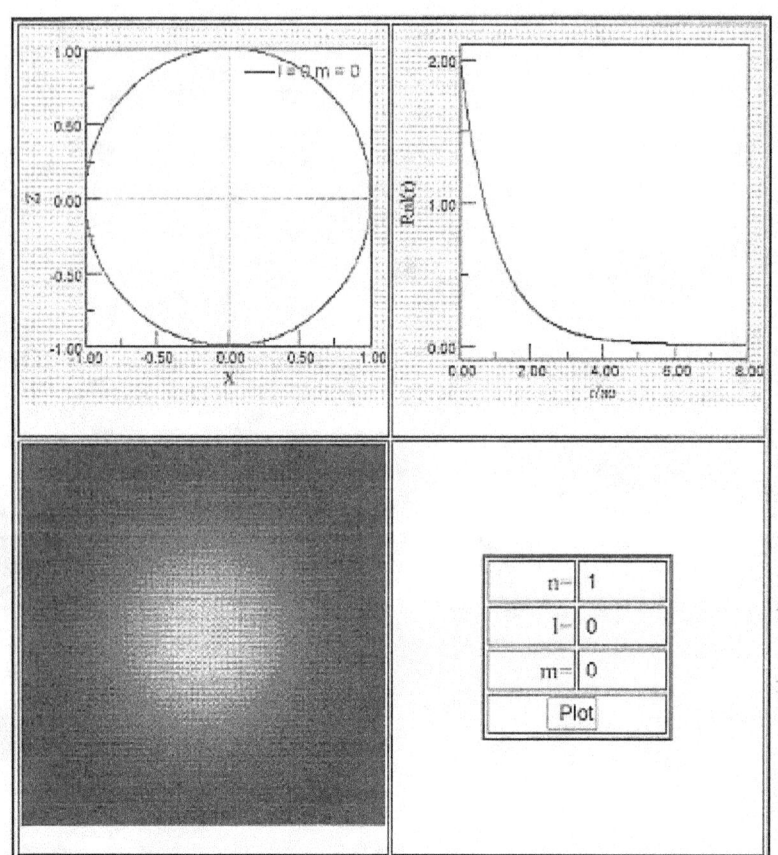

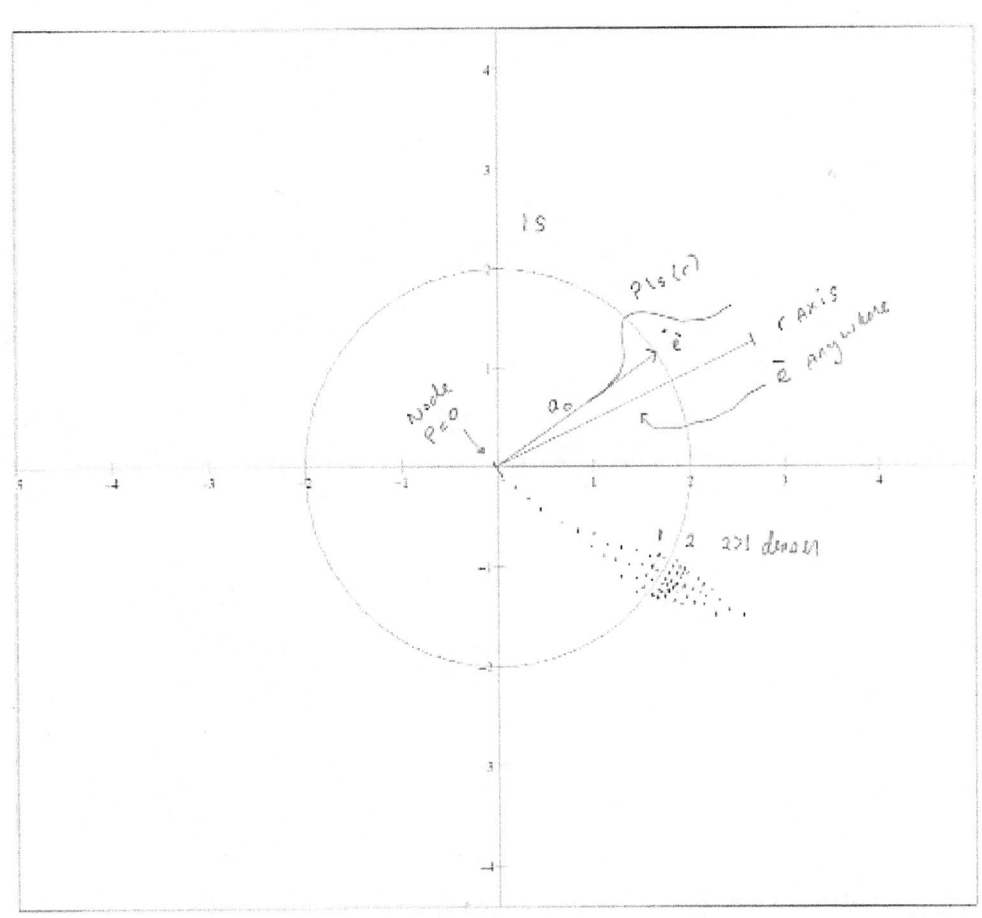

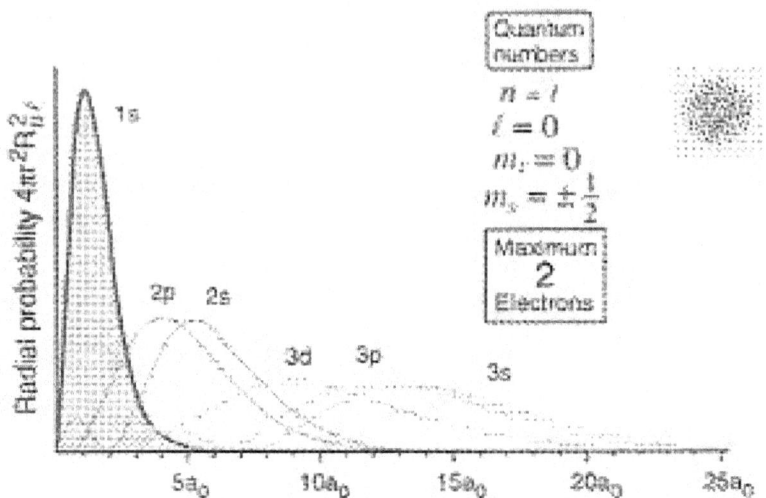

Notes:

2s state: 1st excited state $\Psi n,l,ml\ (r,\theta,\phi)$
$\quad\quad\quad\quad\quad\quad\quad\quad\quad\quad\quad\quad$ 2 0 0
$\quad\quad\quad\quad\quad\quad\quad\quad\quad\quad\quad\quad$ R P F

#1: $\quad e^{-r} - r \cdot e^{-r}$

Normalized and constants eliminated.

$[1/\sqrt{(2\pi)}]\ [1/\sqrt{2}]\ [(1/[2*\sqrt{2}*a(0)^\wedge(3/2)]]*(2-r/a(0))*e^\wedge(-r/a(0))$
$\ F \quad\quad\quad\quad P \quad\quad\quad R$
$\ c \quad\quad\quad\quad c$

ψ2s single variable is r.
Spherical
E2= -3.401 eV
Most probable value is the hight peak = P(5*a(0)) = .265 nm
Electron is farther from the nucleus.

#2: $\quad \left| e^{-r} - r \cdot e^{-r} \right|^2$

#3: $\quad\quad\quad\quad\quad\quad\quad\quad\quad\quad e^{-2 \cdot r} \cdot (r - 1)^2$

Probability density

#4: $\quad (e^{-2 \cdot r} \cdot (r - 1)^2) \cdot (4 \cdot \pi \cdot r^2)$

$\quad\quad\quad\quad\quad\quad\quad\quad$ shell volume

#5: $\quad\quad\quad\quad\quad\quad\quad\quad 12.56637061 \cdot r^2 \cdot e^{-2 \cdot r} \cdot (r - 1)^2$

Radial density function

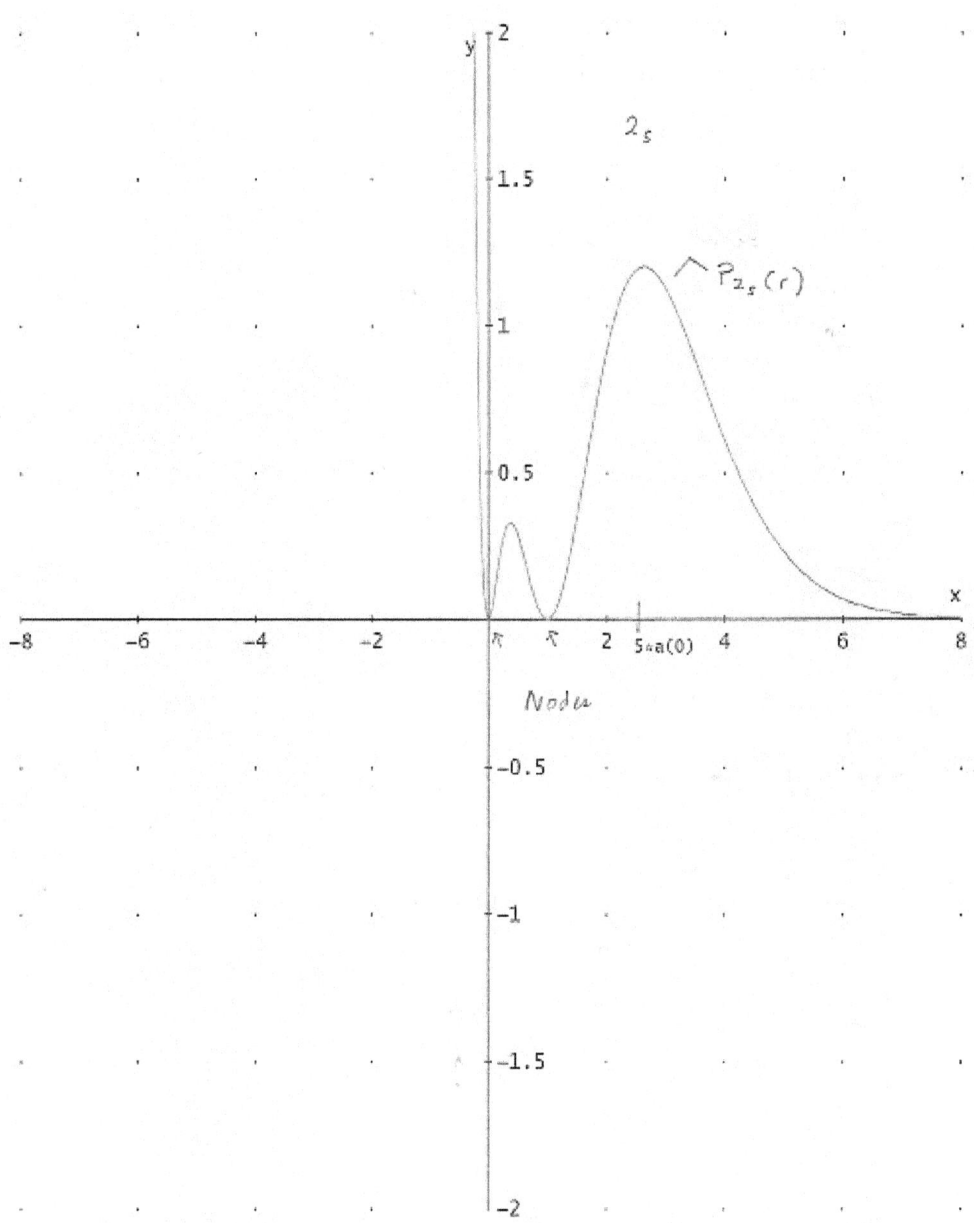

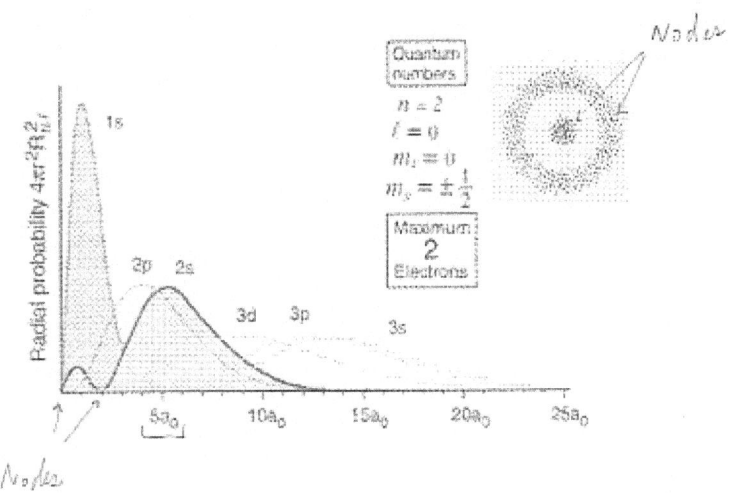

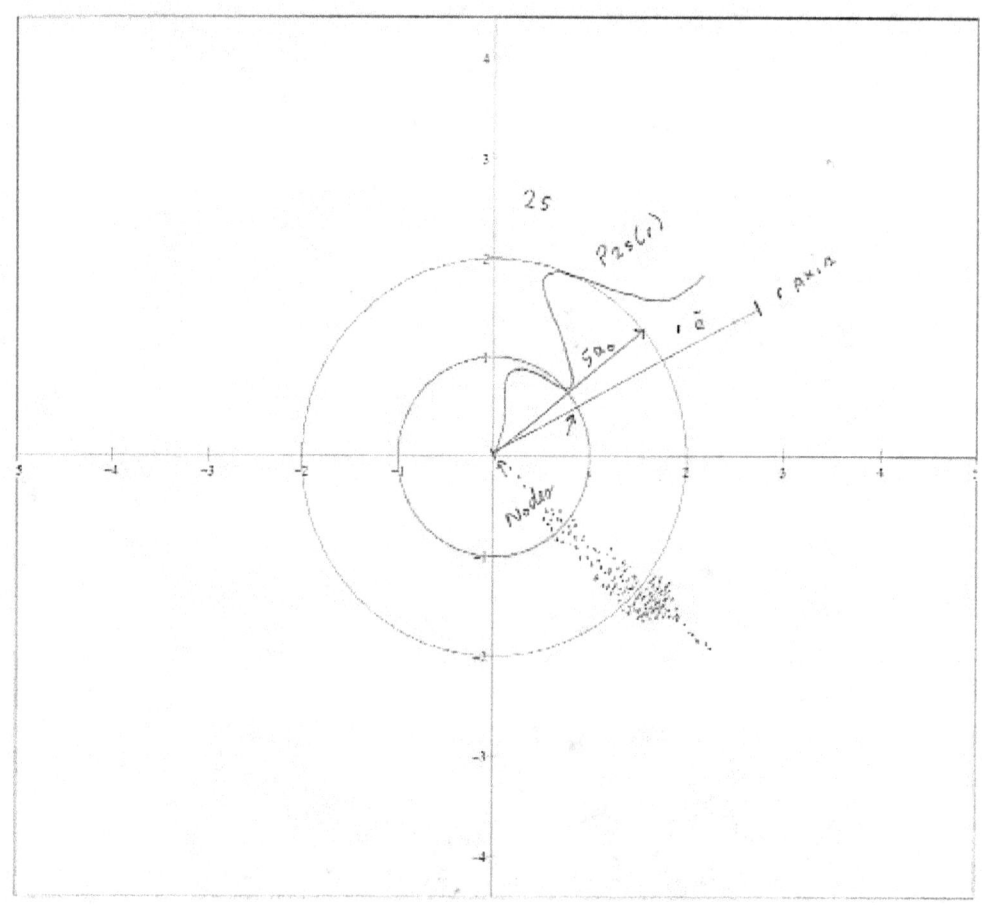

3s state

#1: $$\frac{1}{\sqrt{2\cdot\pi}}\cdot\frac{1}{\sqrt{2}}\cdot\frac{2}{81\cdot\sqrt{3}\cdot a^{3/2}}\cdot\left(\left(27-18\cdot\frac{r}{a}+\frac{2\cdot r^2}{a^2}\right)\cdot e^{-r/(3\cdot a)}\right)$$

F   P   R

Normalized

#2: $$\frac{10000\cdot\sqrt{1590}\cdot e^{-1000\cdot r/159}\cdot(2000000\cdot r^2-954000\cdot r+75843)}{1917386883\cdot\sqrt{\pi}}$$

a=.053 nm

#3: $$\left|\frac{1}{\sqrt{2\cdot\pi}}\cdot\frac{1}{\sqrt{2}}\cdot\frac{2}{81\cdot\sqrt{3}\cdot a^{3/2}}\cdot\left(\left(27-18\cdot\frac{r}{a}+\frac{2\cdot r^2}{a^2}\right)\cdot e^{-r/(3\cdot a)}\right)\right|^2$$

density

#4: $$\frac{1000000000\cdot e^{-2000\cdot r/159}\cdot(2000000\cdot r^2-954000\cdot r+75843)^2}{23121839365411671\cdot\pi}$$

a=.053 nm

#5: $$\left|\frac{1}{\sqrt{2\cdot\pi}}\cdot\frac{1}{\sqrt{2}}\cdot\frac{2}{81\cdot\sqrt{3}\cdot a^{3/2}}\cdot\left(\left(27-18\cdot\frac{r}{a}+\frac{2\cdot r^2}{a^2}\right)\cdot e^{-r/(3\cdot a)}\right)\right|^2\cdot(4\cdot\pi\cdot r^2)$$

density function

#6: $$\frac{4000000000\cdot r^2\cdot e^{-2000\cdot r/159}\cdot(2000000\cdot r^2-954000\cdot r+75843)^2}{23121839365411671}$$

a=.053 nm

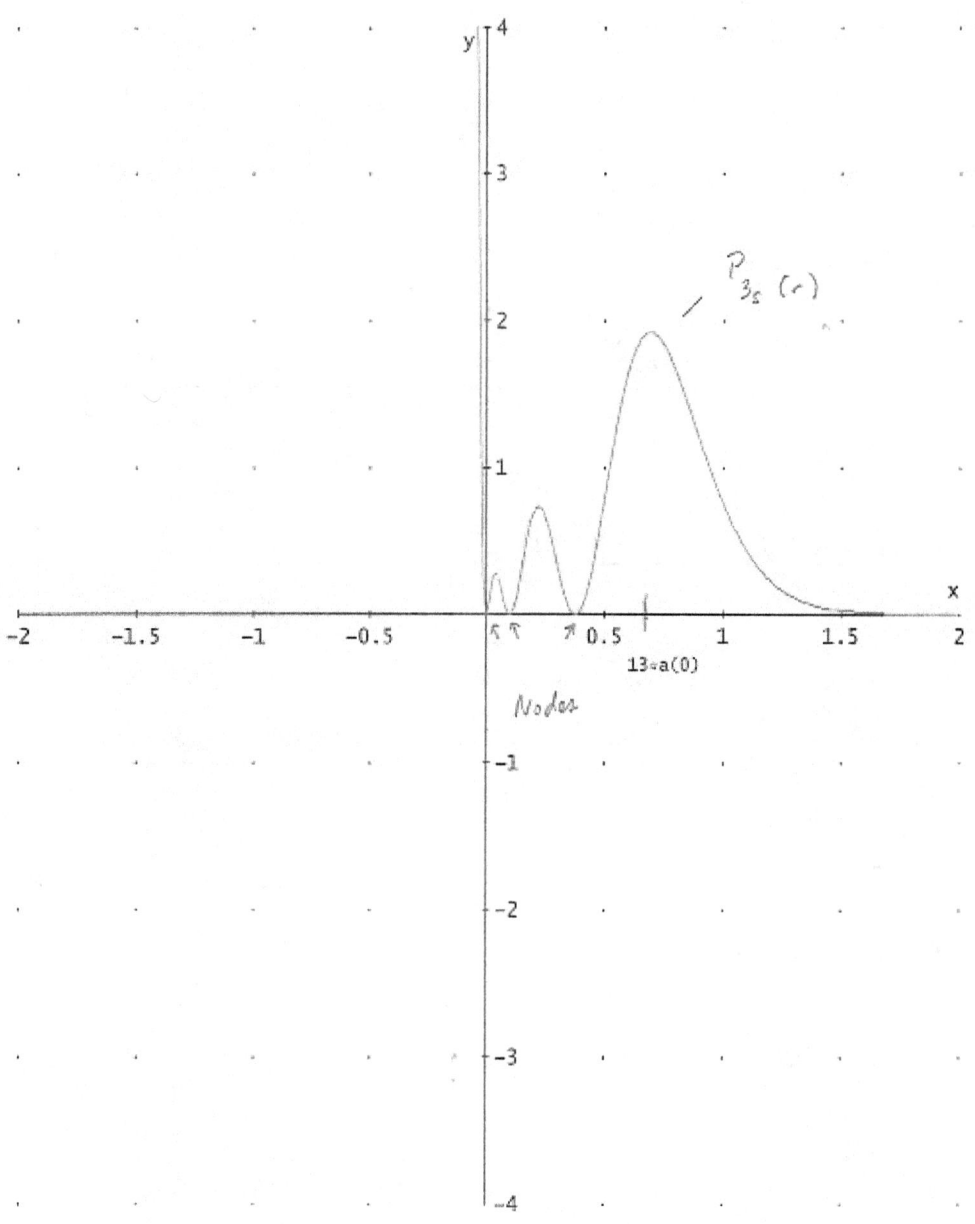

$P_{3s}(r)$

Nodes

13=a(0)

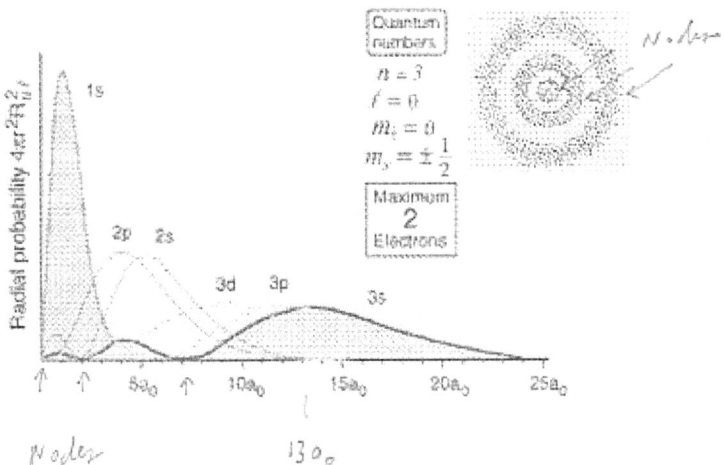

Nodes

13 a_0

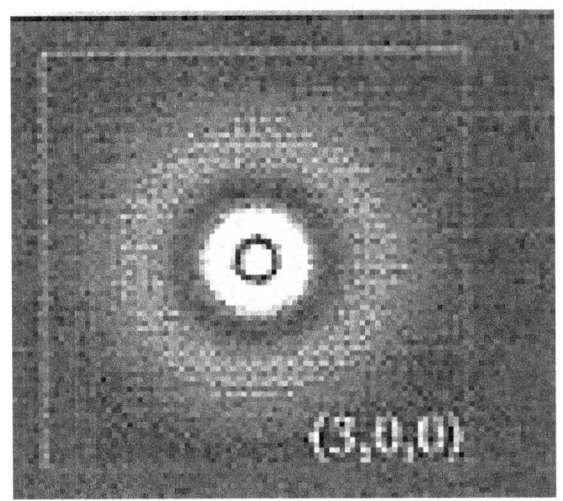

(3,0,0)

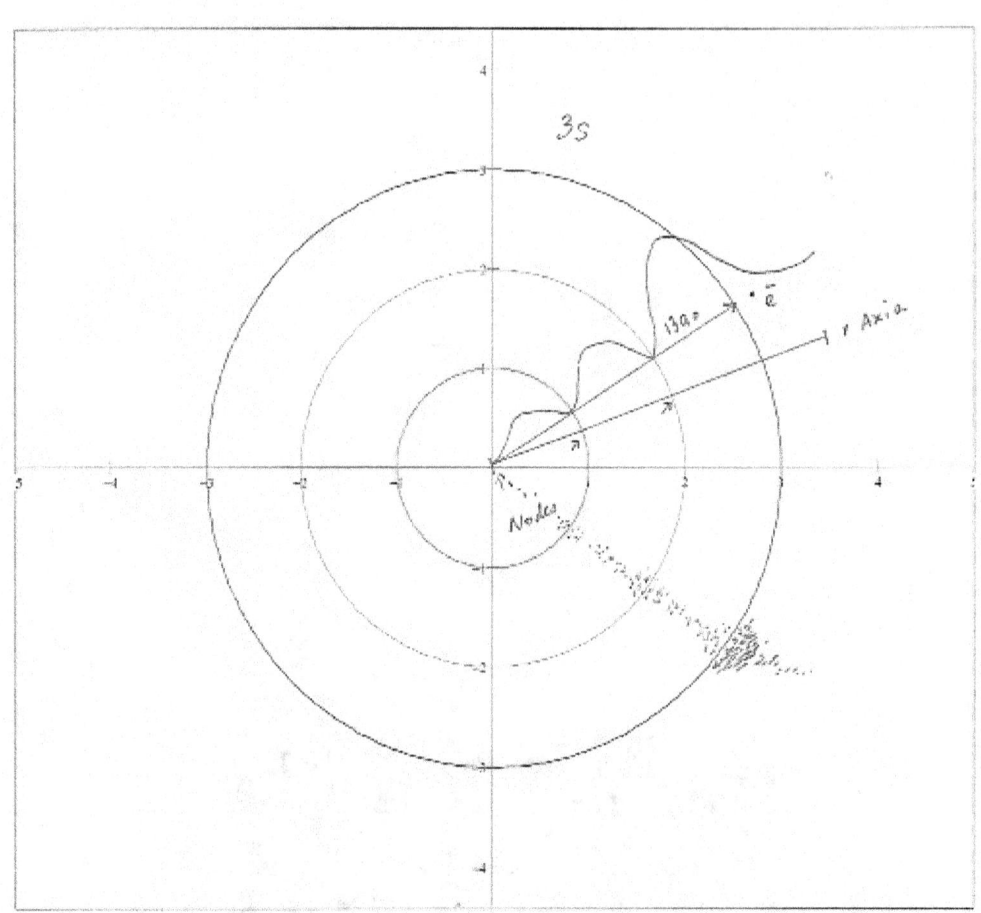

```
2p Orbital
n l ml
2 1 0
F(φ)  P(θ)   R(r)
c     cosθ   r*e^(-r)
```

#1: $\dfrac{1}{\sqrt{(2\cdot\pi)}}$

#2: $\dfrac{1}{\sqrt{(2\cdot\pi)}} \cdot \dfrac{\sqrt{6}}{6} \cdot \cos(\theta)$

#3: $\dfrac{1}{\sqrt{(2\cdot\pi)}} \cdot \dfrac{\sqrt{6}}{6} \cdot \cos(\theta) \cdot \dfrac{1}{(2\cdot\sqrt{6})\cdot a^{3/2}}$

#4: $\dfrac{1}{\sqrt{(2\cdot\pi)}} \cdot \dfrac{\sqrt{6}}{6} \cdot \cos(\theta) \cdot \dfrac{1}{(2\cdot\sqrt{6})\cdot a^{3/2}} \cdot \dfrac{r}{a}$

#5: $\dfrac{1}{\sqrt{(2\cdot\pi)}} \cdot \dfrac{\sqrt{6}}{6} \cdot \cos(\theta) \cdot \dfrac{1}{(2\cdot\sqrt{6})\cdot a^{3/2}} \cdot \dfrac{r}{a} \cdot e^{-r/(2\cdot a)}$

Normalized Ψ

#6: $\dfrac{0.03324519003\cdot r \cdot e^{-0.5\cdot r/a} \cdot \cos(\theta)}{a^{2.5}}$

#7: $\dfrac{0.03324519003\cdot r \cdot e^{-0.5\cdot r/0.053} \cdot \cos(\theta)}{0.053^{2.5}}$

a=0.053

#8: $-\dfrac{0.03324519003\cdot r \cdot e^{-0.5\cdot r/0.053} \cdot \cos(\theta)}{0.053^{2.5}}$

#9: $\left| \dfrac{0.03324519003 \cdot r \cdot e^{-0.5 \cdot r / 0.053} \cdot \cos(\theta)}{0.053^{2.5}} \right|$

#10: $\left| \dfrac{0.03324519003 \cdot r \cdot e^{-0.5 \cdot r / 0.053} \cdot \cos(\theta)}{0.053^{2.5}} \right|^{2}$

Density

#11: $-\left| \dfrac{0.03324519003 \cdot r \cdot e^{-0.5 \cdot r / 0.053} \cdot \cos(\theta)}{0.053^{2.5}} \right|^{2}$

Notes:

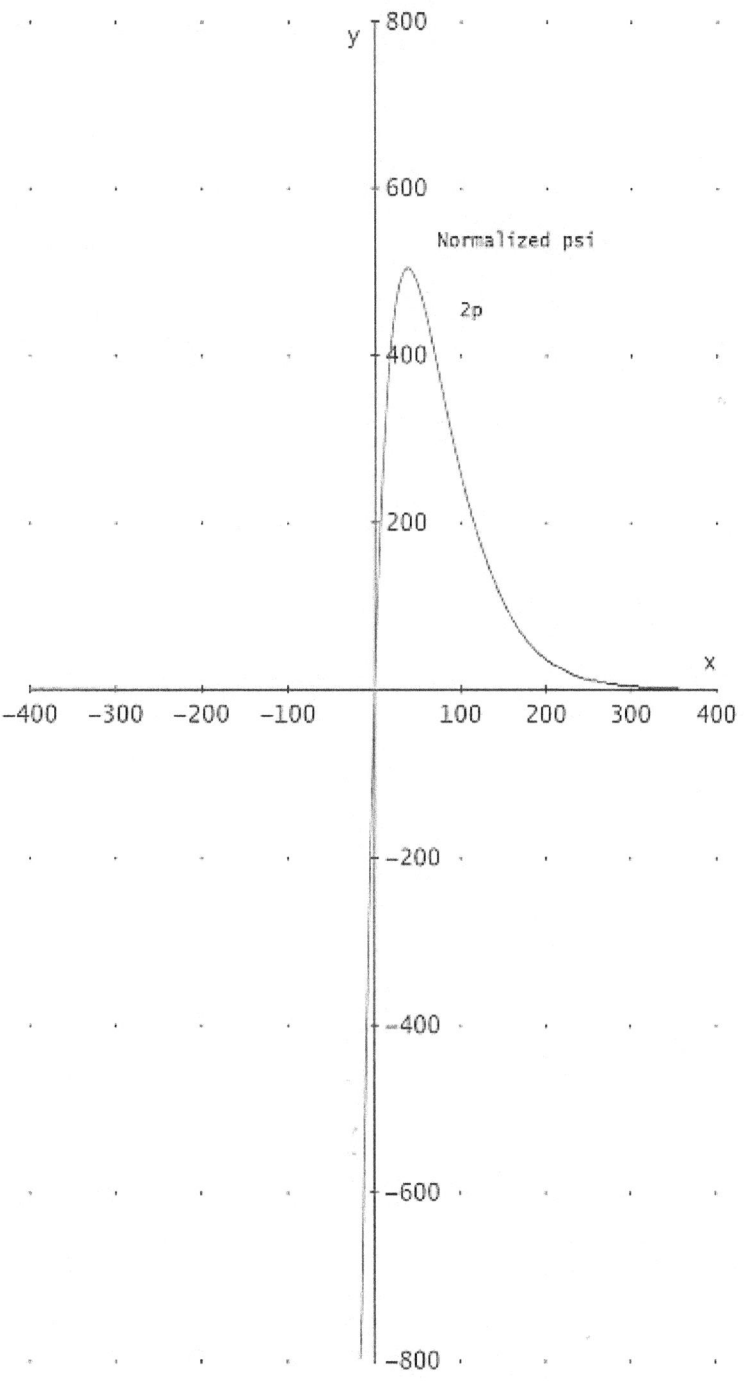

65

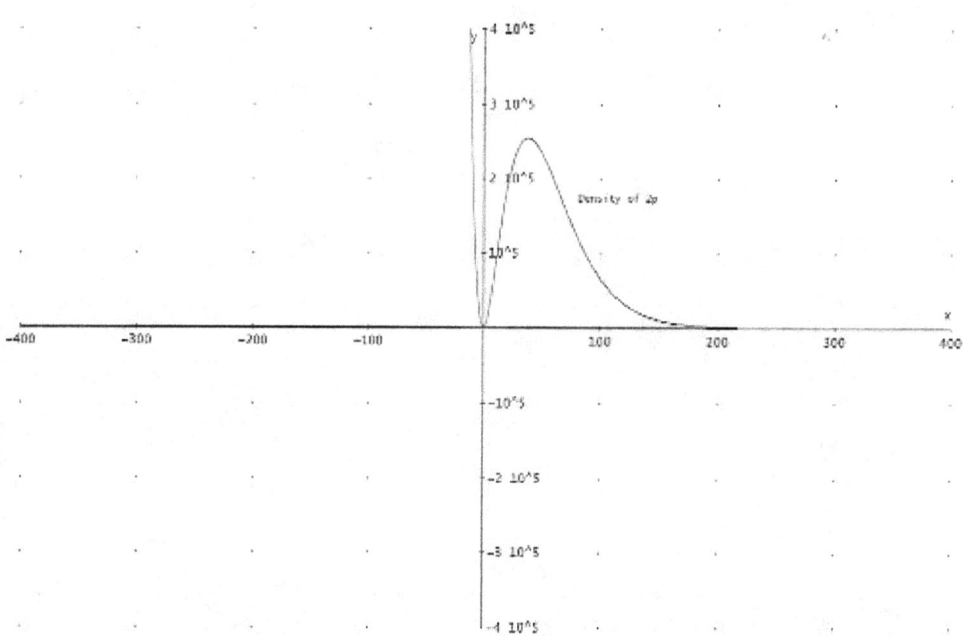

Notes:

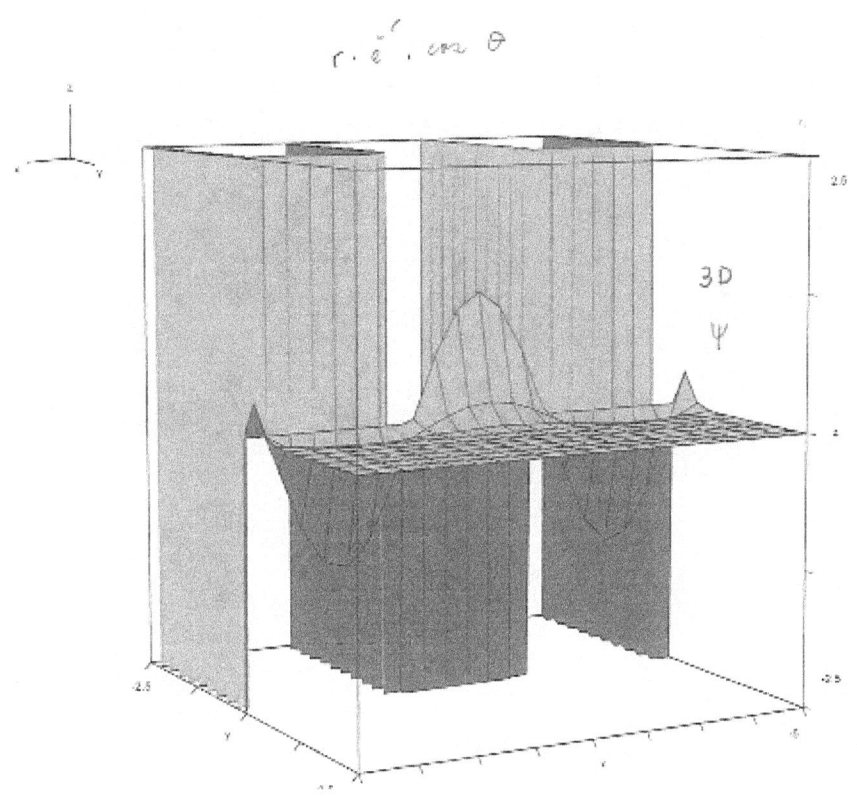

$r \cdot \bar{e}^{r} \cdot \cos \theta$

$\sim r \cdot e^{-r} \cdot \cos\theta$

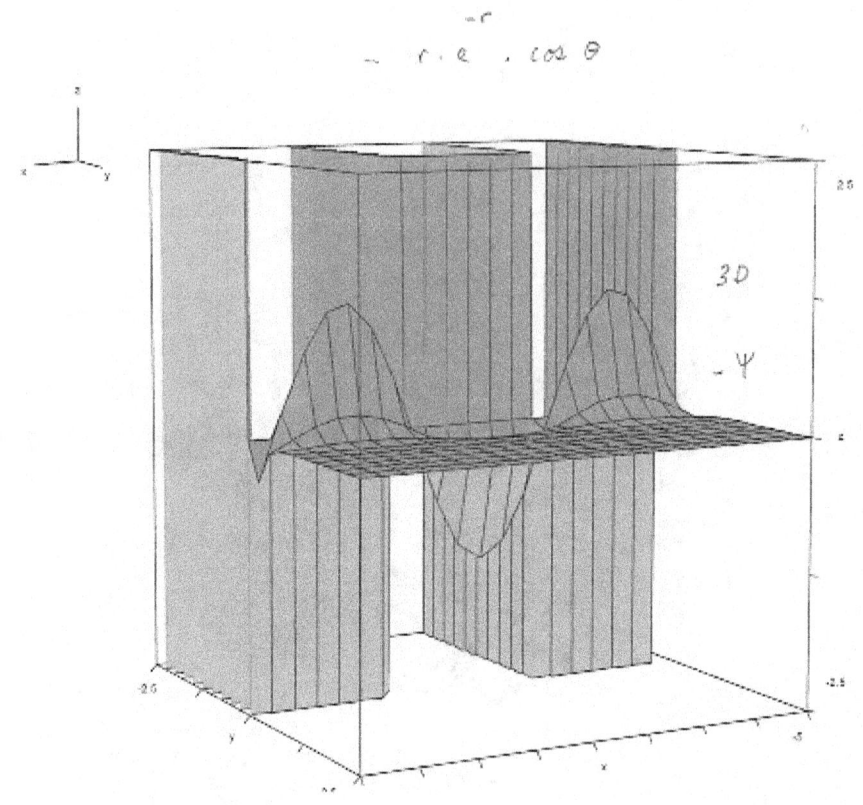

3D

$\sim \Psi$

Superposition

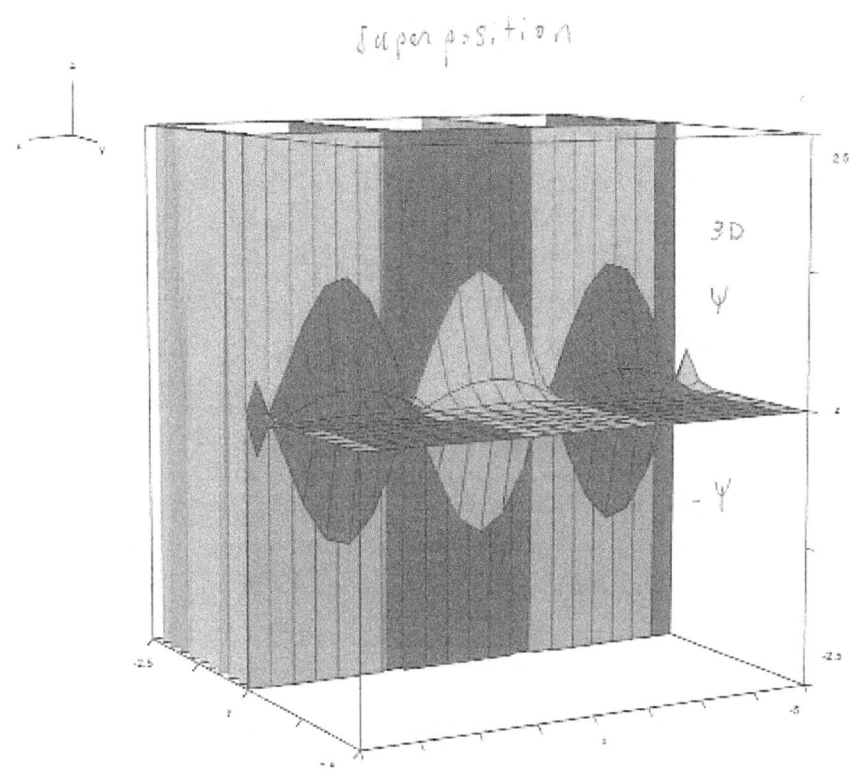

3D

ψ

-ψ

Superposition

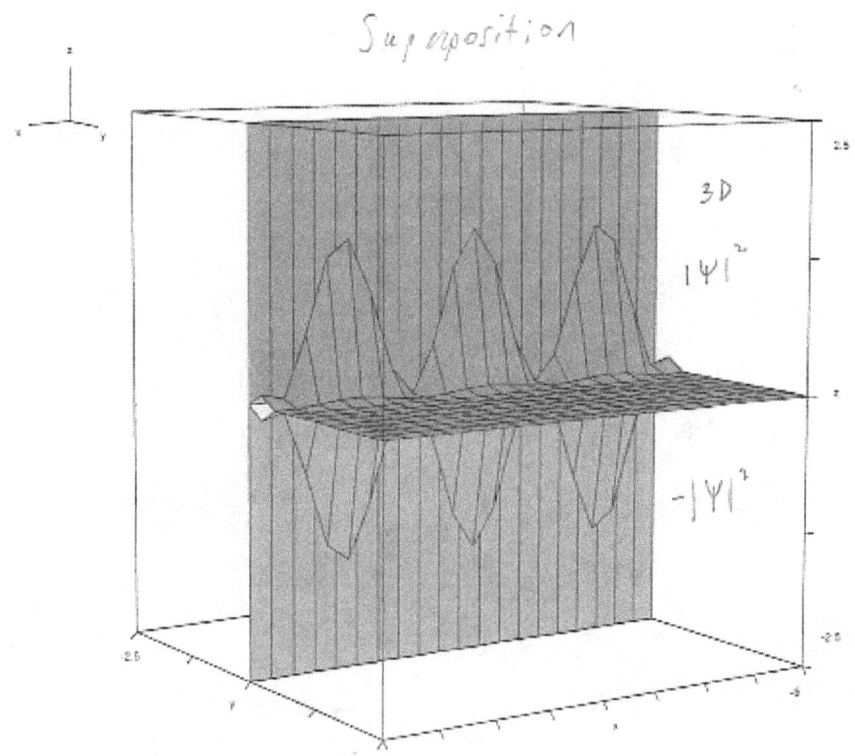

3D

$|\Psi|^2$

$-|\Psi|^2$

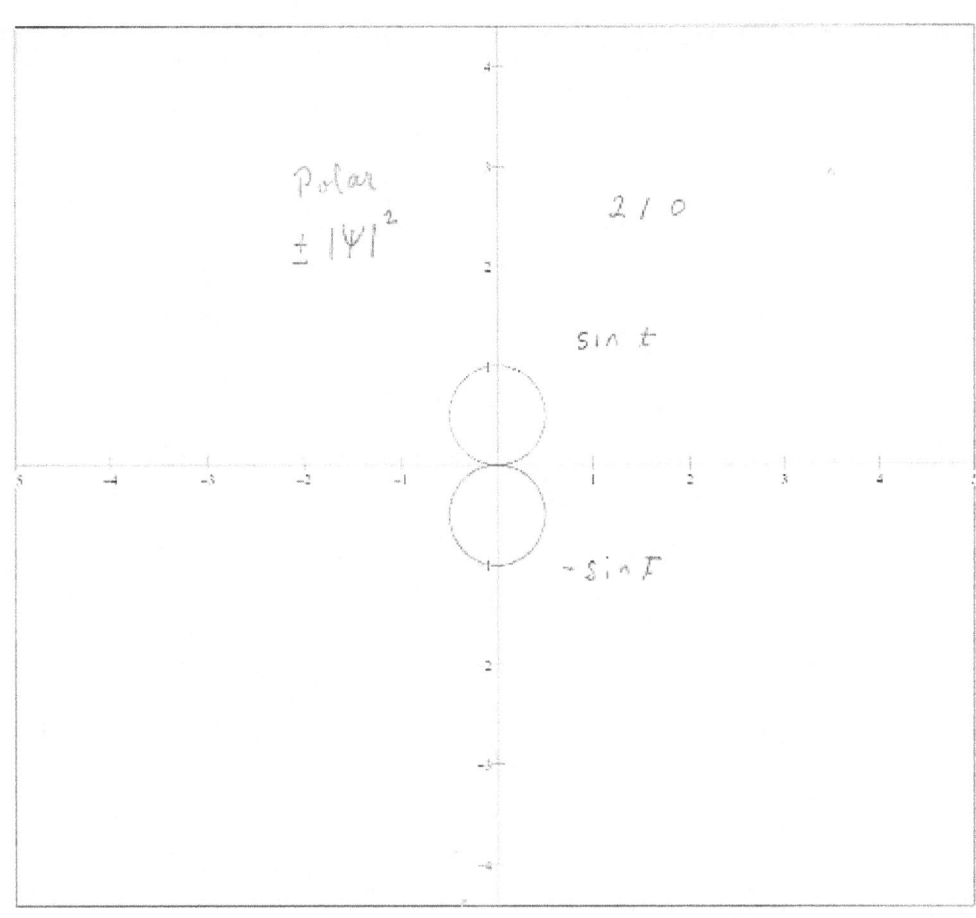

Polar
$\pm |\Psi|^2$

$2 / 0$

sin t

$-\sin t$

(2,1,0)

#2 2p orbit (+i) 2 1 +1 Elliptical

#1:
$$\frac{1}{8\cdot\sqrt{\pi}\cdot a^{3/2}}\cdot\frac{r}{a}\cdot e^{-r/(2\cdot a)}\cdot SIN(\theta)\cdot e^{i\cdot\phi}$$

#2:
$$\frac{1250000\cdot\sqrt{265}\cdot r\cdot e^{-500\cdot r/53 + i\cdot\phi}}{148877\cdot\sqrt{\pi}}$$

#3:
$$\frac{625000\cdot\sqrt{530}\cdot r\cdot e^{-500\cdot r/53}}{148877\cdot\sqrt{\pi}} + \frac{625000\cdot\sqrt{530}\cdot i\cdot r\cdot e^{-500\cdot r/53}}{148877\cdot\sqrt{\pi}}$$

substituted: θ & ϕ = $\pi/4$ (+) Imaginary and (+)real.superimposed.

#4:
$$\left|\frac{625000\cdot\sqrt{530}\cdot r\cdot e^{-500\cdot r/53}}{148877\cdot\sqrt{\pi}} + \frac{625000\cdot\sqrt{530}\cdot i\cdot r\cdot e^{-500\cdot r/53}}{148877\cdot\sqrt{\pi}}\right|^{2}$$

Density

#3 2p orbit polar opposite (-i) 2 1 -1 Elliptical.

#5:
$$\frac{1}{8\cdot\sqrt{\pi}\cdot a^{3/2}}\cdot\frac{r}{a}\cdot e^{-r/(2\cdot a)}\cdot SIN(\theta)\cdot e^{-i\cdot\phi}$$

#6:
$$\frac{625000\cdot\sqrt{530}\cdot r\cdot e^{-500\cdot r/53}}{148877\cdot\sqrt{\pi}} - \frac{625000\cdot\sqrt{530}\cdot i\cdot r\cdot e^{-500\cdot r/53}}{148877\cdot\sqrt{\pi}}$$

substituted: (+) real and (-) Imaginary. Non-superimposed.

#7:
$$\left|\frac{625000\cdot\sqrt{530}\cdot r\cdot e^{-500\cdot r/53}}{148877\cdot\sqrt{\pi}} - \frac{625000\cdot\sqrt{530}\cdot i\cdot r\cdot e^{-500\cdot r/53}}{148877\cdot\sqrt{\pi}}\right|^{2}$$

#8:
$$-\left|\frac{625000\cdot\sqrt{530}\cdot r\cdot e^{-500\cdot r/53}}{148877\cdot\sqrt{\pi}} - \frac{625000\cdot\sqrt{530}\cdot i\cdot r\cdot e^{-500\cdot r/53}}{148877\cdot\sqrt{\pi}}\right|^{2}$$

Polar opposite density: (-i).

There are 3 2p orbitals: $\cos\theta$, $\sin\theta*e^{\wedge}(i\phi)$, $\sin\theta*e^{\wedge}(-i\phi)$. All down the 3 axes: x,y, and z. Making 90° to each other, orthogonal.

Only the real values make up the 3 orbits.

The states: 2 0 0 Sphere
2 1 0 Ellipse
2 1 +1 Ellipse
2 1 -1 Ellipse

Notes:

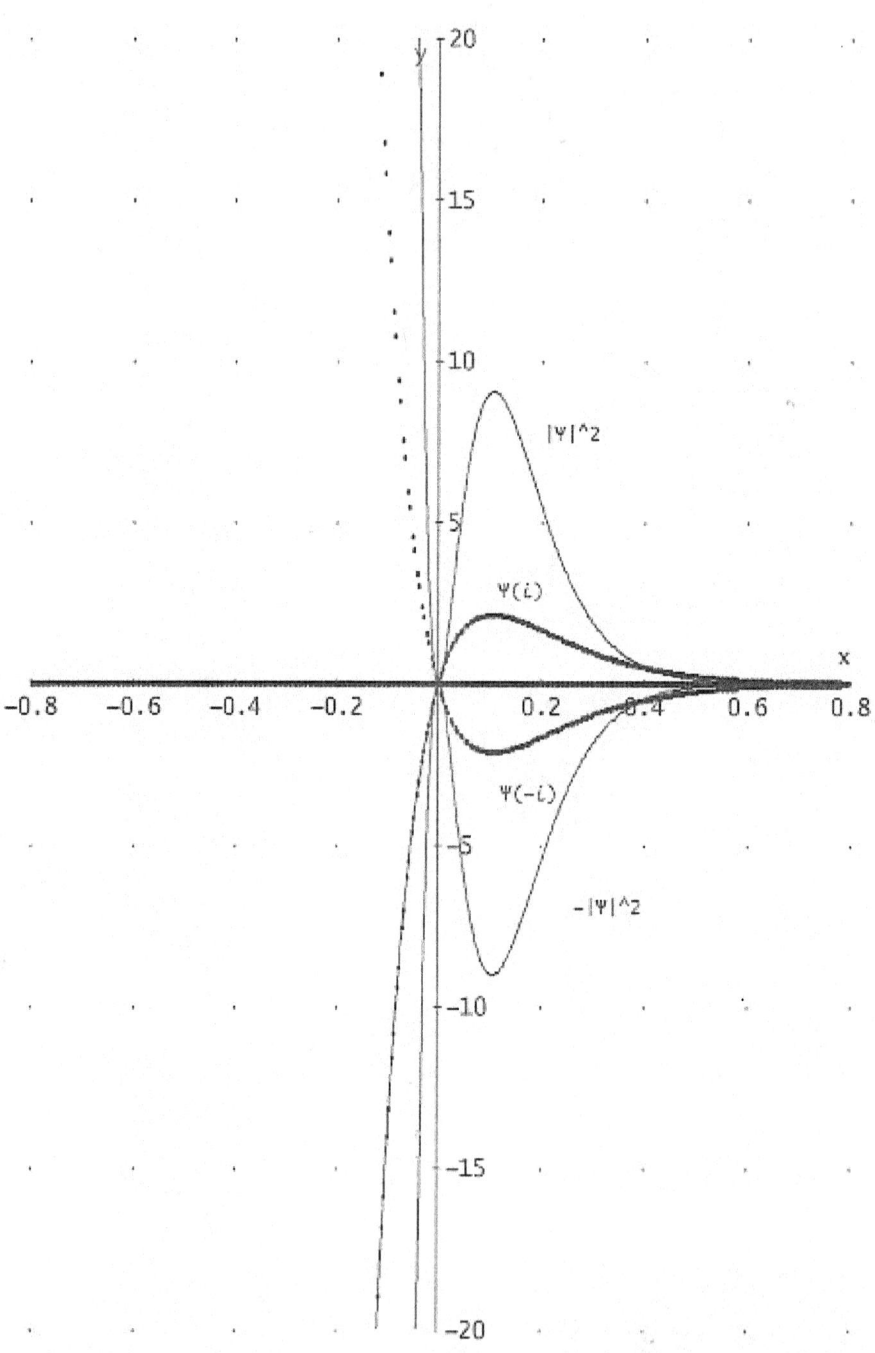

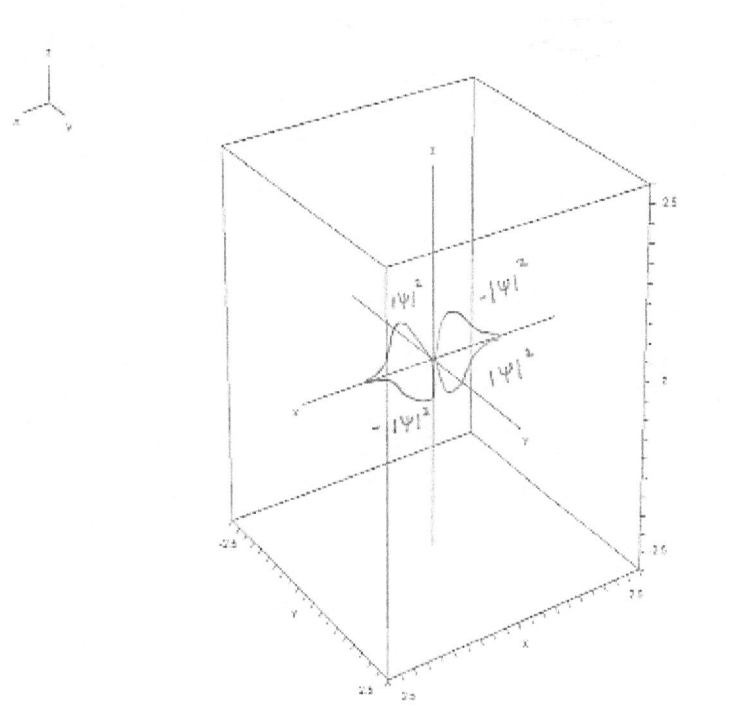

$|\Psi|^2$

Ellipse

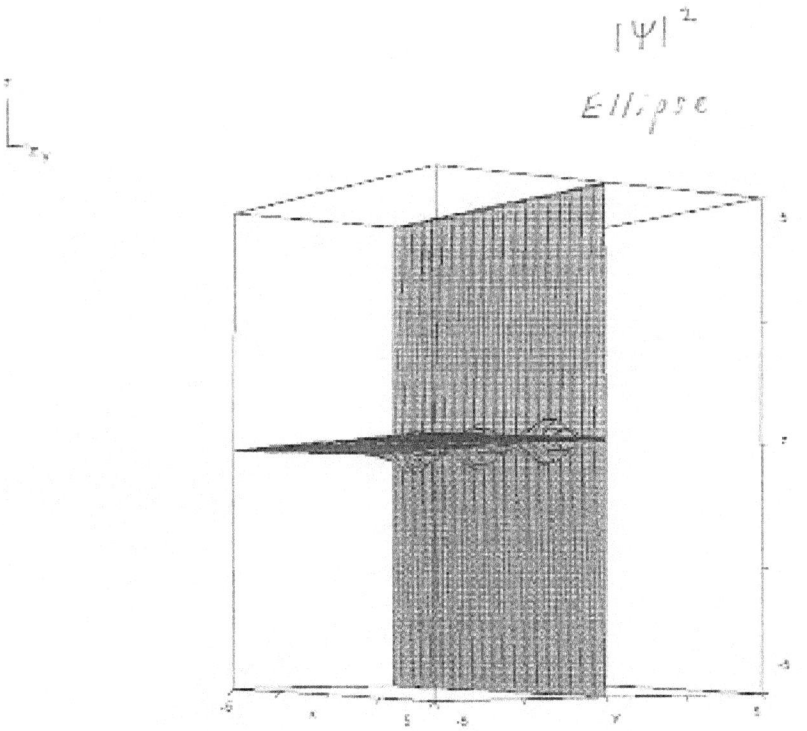

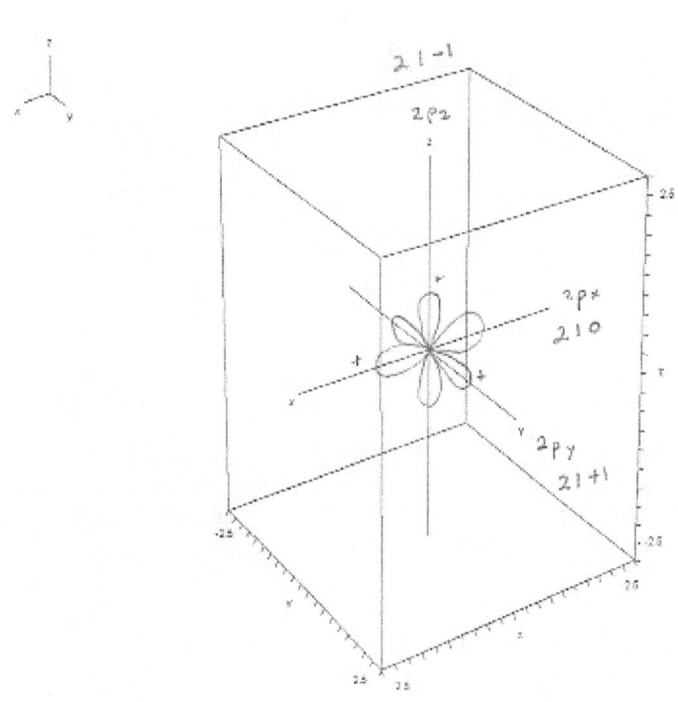

Some basic trigonometry: cos = sin (just out of phase by 90 degrees).
Multiplying by (-1) gives the polar opposite of function.
Changing the independent variable changes axis.

#1: $\cos\left(x + \dfrac{\pi}{2}\right)$

#2: $-\cos\left(x + \dfrac{\pi}{2}\right)$

#3: $\text{SIN}(y + \pi)$

#4: $-\text{SIN}(y + \pi)$

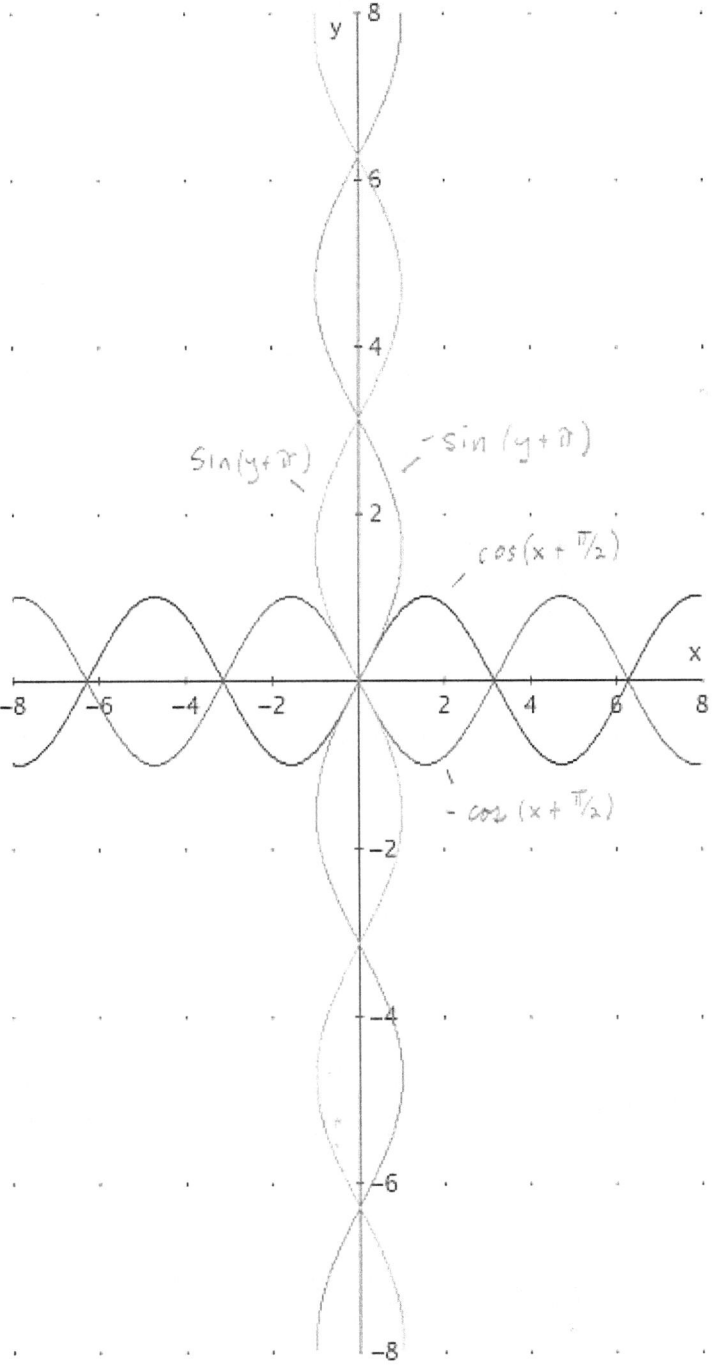

$\sin(y+\pi)$ $-\sin(y+\pi)$

$\cos(x+\frac{\pi}{2})$

$-\cos(x+\frac{\pi}{2})$

$\cos x \qquad \cos\left(x - \frac{\pi}{2}\right) = \sin x$

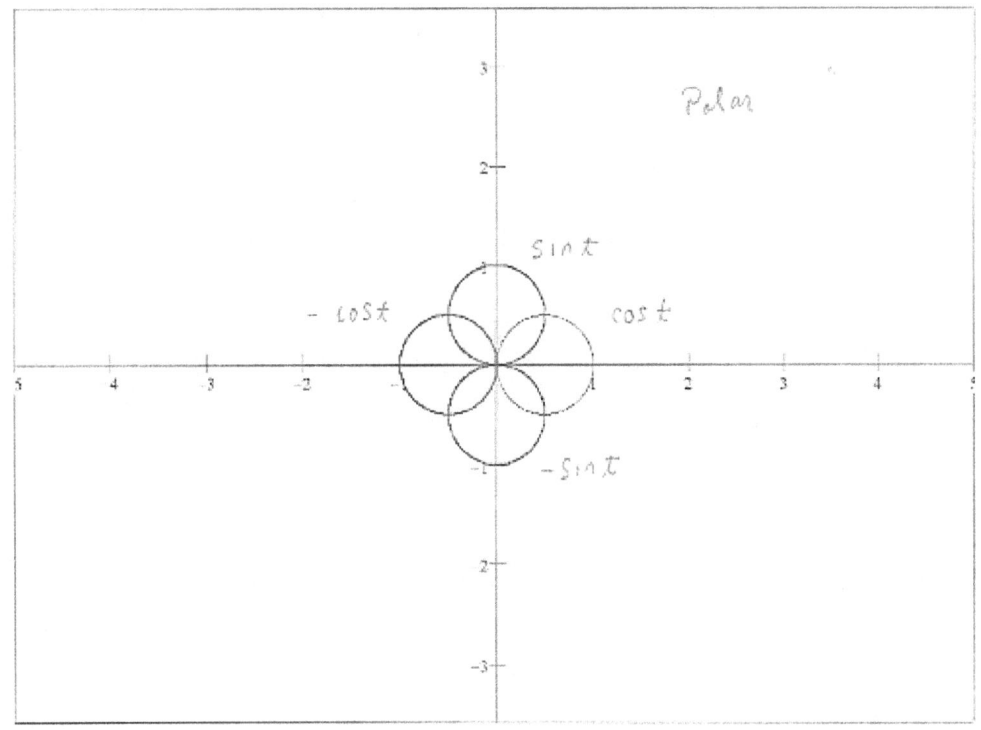

(2.1.1)

3p : 3 1 0

#1: $\dfrac{\sqrt{2}}{81\cdot\sqrt{\pi}\cdot a^{3/2}}\cdot\left(6-\dfrac{r}{a}\right)\cdot\dfrac{r}{a}\cdot e^{-r/(3\cdot a)}\cdot\cos(\theta)$

#2: $\dfrac{\sqrt{2}}{81\cdot\sqrt{\pi}\cdot 0.053^{3/2}}\cdot\left(6-\dfrac{r}{0.053}\right)\cdot\dfrac{r}{0.053}\cdot e^{-r/(3\cdot 0.053)}\cdot\cos\left(\dfrac{\pi}{4}\right)$

#3: $\left|\dfrac{\sqrt{2}}{81\cdot\sqrt{\pi}\cdot 0.053^{3/2}}\cdot\left(6-\dfrac{r}{0.053}\right)\cdot\dfrac{r}{0.053}\cdot e^{-r/(3\cdot 0.053)}\cdot\cos\left(\dfrac{\pi}{4}\right)\right|^{2}$

3p 3 1 1

#4: $\dfrac{1}{81\cdot\sqrt{\pi}\cdot a^{3/2}}\cdot\left(6-\dfrac{r}{a}\right)\cdot\dfrac{r}{a}\cdot e^{-r/(3\cdot a)}\cdot\sin(\theta)\cdot e^{i\cdot\phi}$

#5: $\dfrac{10000000\cdot\sqrt{530}\cdot r\cdot e^{-1000\cdot r/159}\cdot(159-500\cdot r)}{639128961\cdot\sqrt{\pi}}+$

$\dfrac{10000000\cdot\sqrt{530}\cdot i\cdot r\cdot e^{-1000\cdot r/159}\cdot(159-500\cdot r)}{639128961\cdot\sqrt{\pi}}$

#6: $\left|\dfrac{10000000\cdot\sqrt{530}\cdot r\cdot e^{-1000\cdot r/159}\cdot(159-500\cdot r)}{639128961\cdot\sqrt{\pi}}+\right.$

$\left.\dfrac{10000000\cdot\sqrt{530}\cdot i\cdot r\cdot e^{-1000\cdot r/159}\cdot(159-500\cdot r)}{639128961\cdot\sqrt{\pi}}\right|^{2}$

3p 3 1 −1

#7: $\dfrac{1}{81\cdot\sqrt{\pi}\cdot a^{3/2}}\cdot\left(6-\dfrac{r}{a}\right)\cdot\dfrac{r}{a}\cdot e^{-r/(3\cdot a)}\cdot\sin(\theta)\cdot e^{-i\cdot\phi}$

$$\#8: \quad \frac{10000000 \cdot \sqrt{530} \cdot r \cdot e^{-1000 \cdot r/159} \cdot (159 - 500 \cdot r)}{639128961 \cdot \sqrt{\pi}} +$$

$$\frac{10000000 \cdot \sqrt{530} \cdot i \cdot r \cdot e^{-1000 \cdot r/159} \cdot (500 \cdot r - 159)}{639128961 \cdot \sqrt{\pi}}$$

$$\#9: \quad \left| \frac{10000000 \cdot \sqrt{530} \cdot r \cdot e^{-1000 \cdot r/159} \cdot (159 - 500 \cdot r)}{639128961 \cdot \sqrt{\pi}} + \frac{10000000 \cdot \sqrt{530} \cdot i \cdot r \cdot e^{-1000 \cdot r/159} \cdot (500 \cdot r - 159)}{639128961 \cdot \sqrt{\pi}} \right|^{2}$$

Note that the 3 1 1 and 3 1 -1 Density is the same; only opposite polarity. If you plot $-i$ in real only, there is no plot. It you plot $-i$ in real and imaginary, there is a plot with polar opposites.

There are 3 3p orbits for a total of 6 electrons: 2 electrons/orbit.

Size: 3>2>1
s orbit is a sphere.
2p orbit is an ellipse.
3p orbit is a sphere and ellipse.

Ψ equation in 3D = $F(\phi) * P(\theta) * R(r)$
In polar function, the waves are symmetrical in the general shape of sin and cos.
$P(\Psi)$ is a multiple of 3 functions.
$|P(\Psi)|^2$ is the absolute squared of $P(\Psi)$.

The orbital shapes are smooth and represent the probability of finding the electron 100% of the time. The difference between $P(\Psi)$ and <x>,mean, is over 50%. Poor in the prediction ability. The uncertainty principle is apparent. The electron is just too fast over a very,very small distance to know "how fast" and "where" for sure.

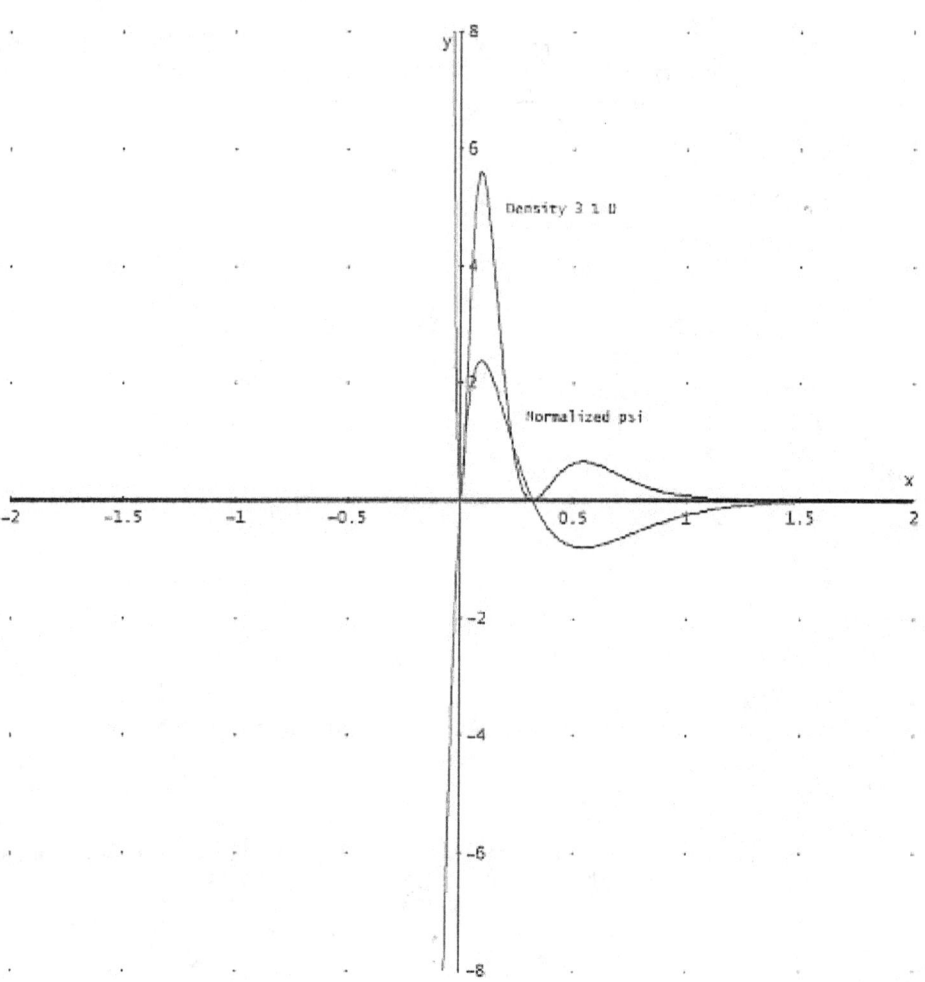

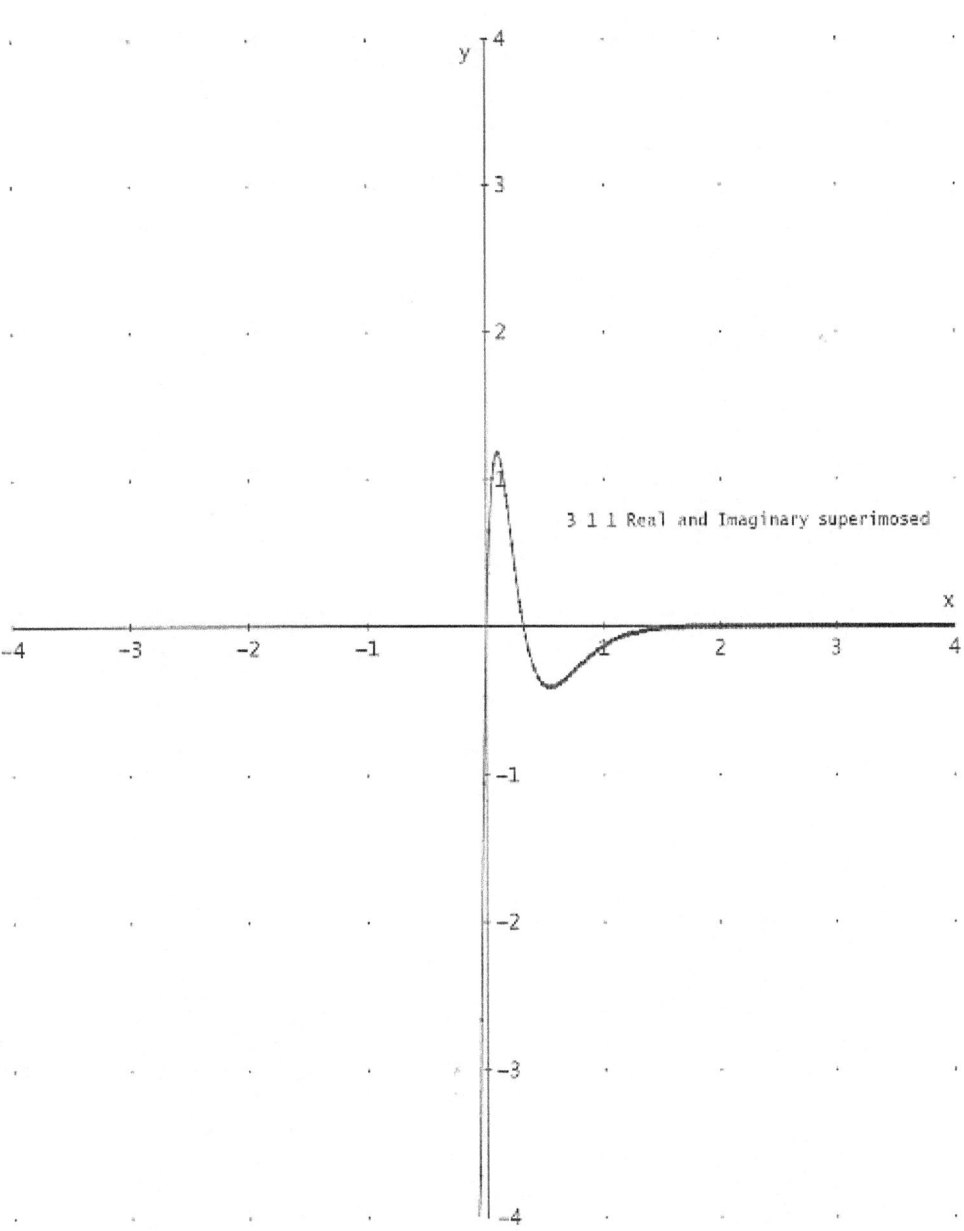

3 1 1 Real and Imaginary superimosed

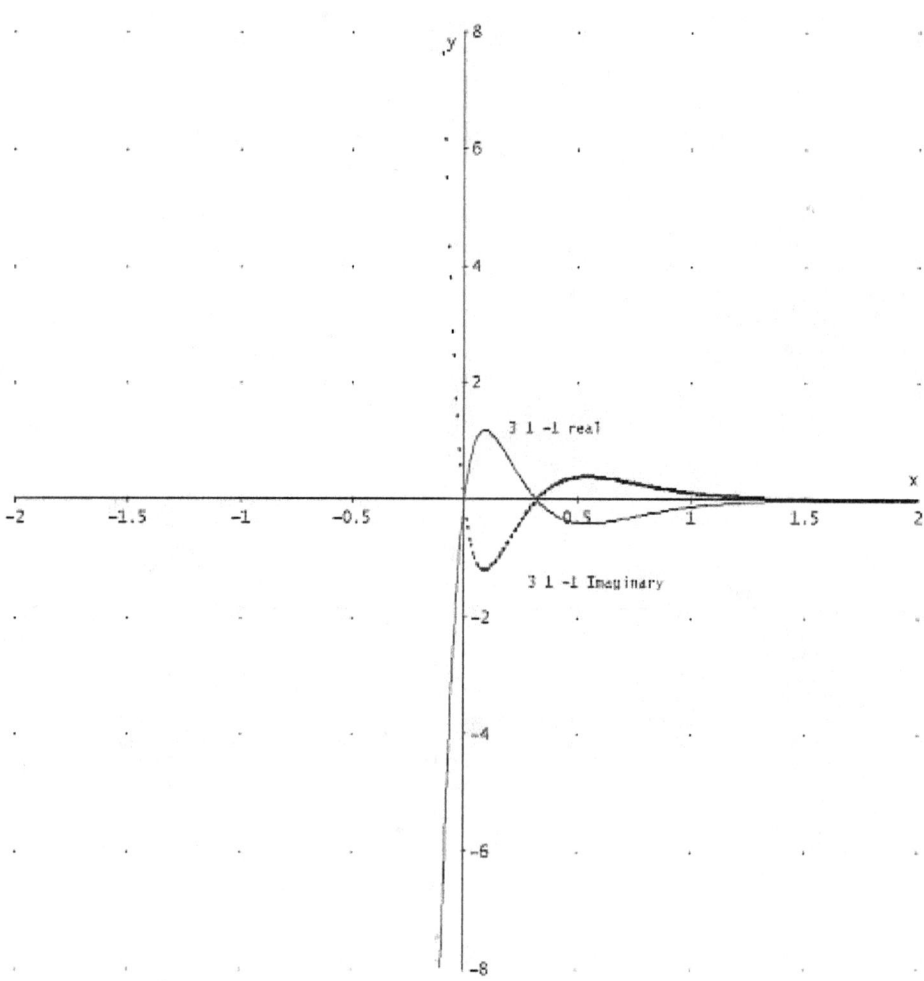

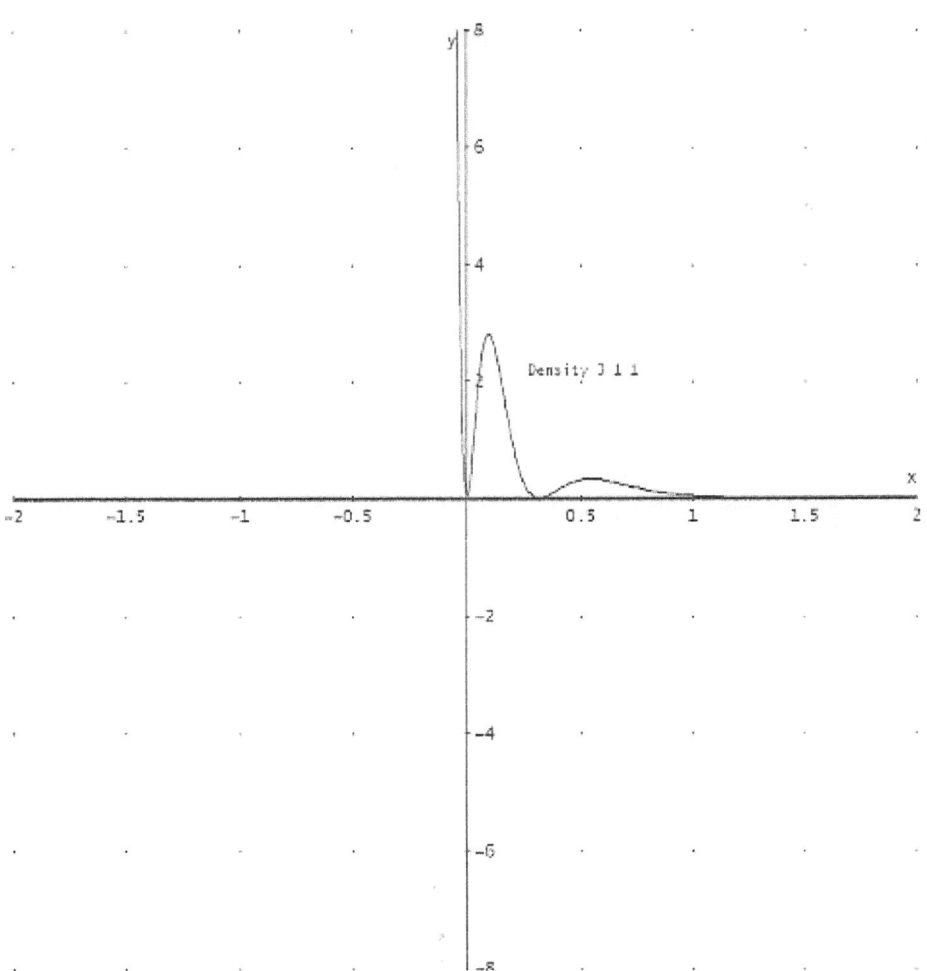

Density 3 1 1

Density 3 1 -1

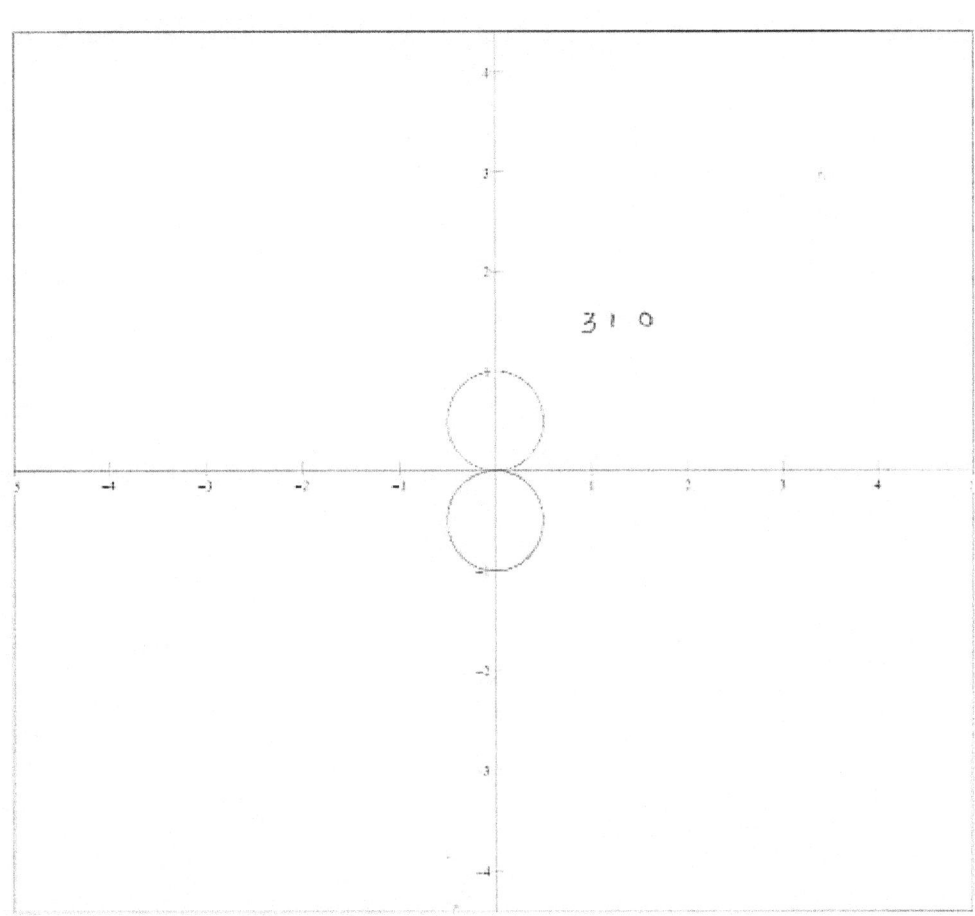

310

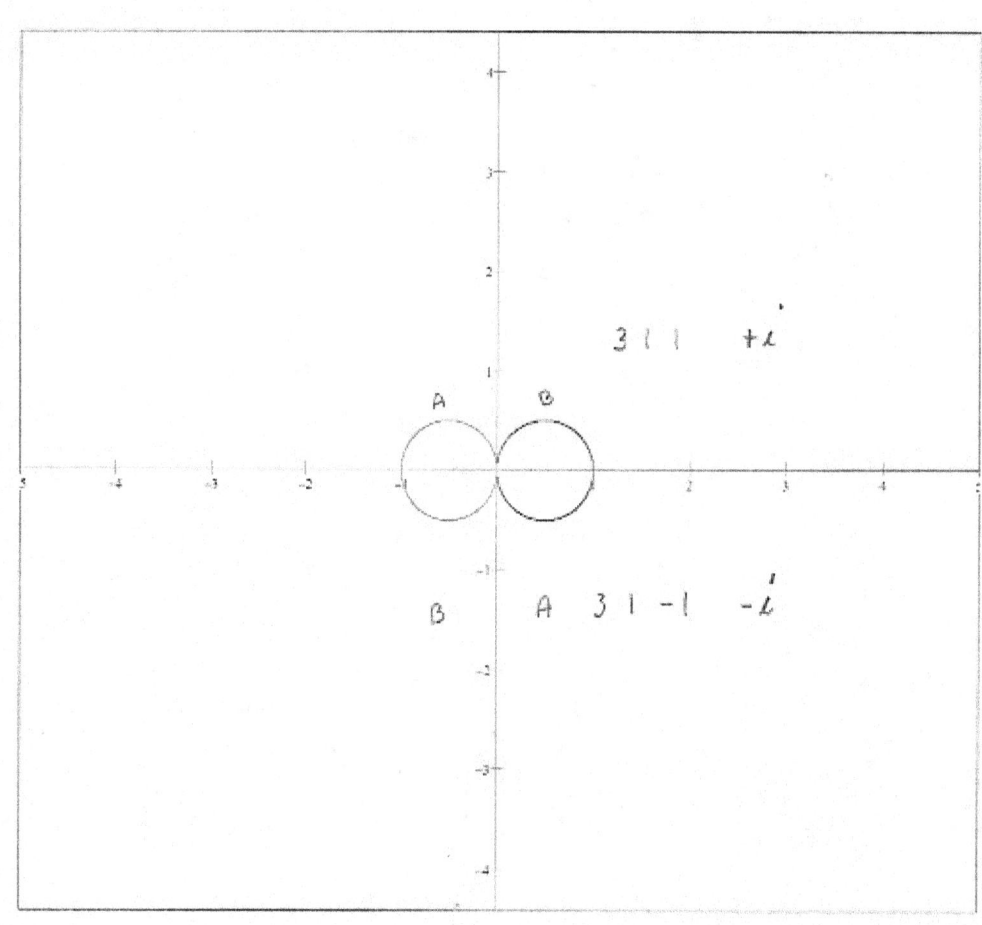

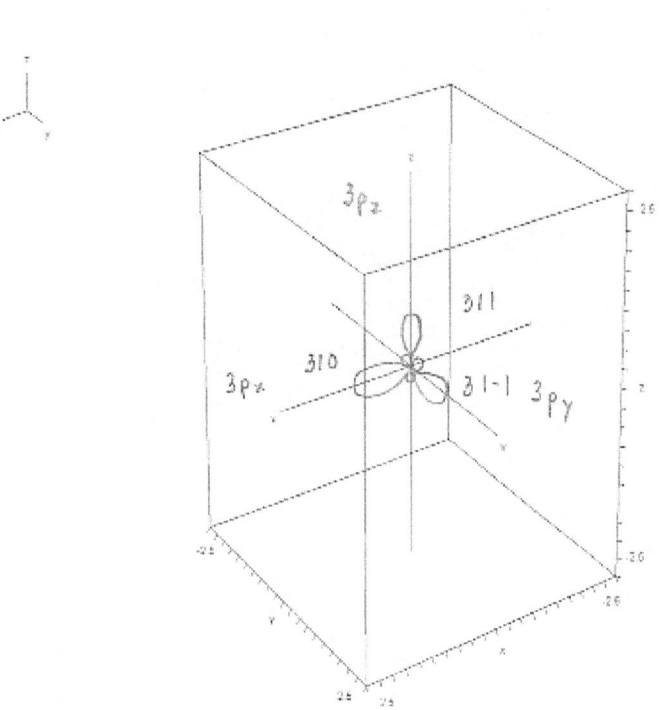

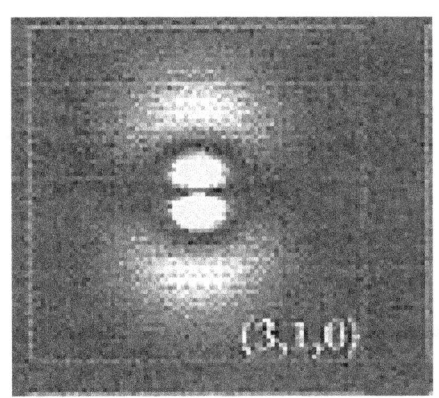

(3,1,0)

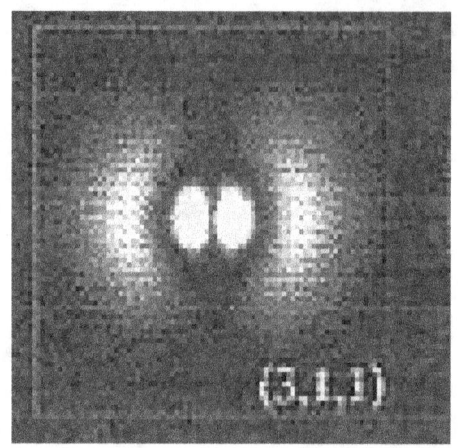

Notes:

Densities

1s

#1: $r^2 \cdot e^{-2\cdot r}$

2s

#2: $r^2 \cdot (r^2 - 2\cdot r + 1)\cdot e^{-2\cdot r}$

2p

#3: $\left| r\cdot e^{-r} \cdot \cos\left(\dfrac{\pi}{4}\right) \right|^2$

2p +/−

#4: $\left| r\cdot e^{-r} + i\cdot r\cdot e^{-r} \right|^2$

#5: $\left| r\cdot e^{-r} - i\cdot r\cdot e^{-r} \right|^2$

Identical : 180° rotated

3s

#6: $\dfrac{(1\cdot 10^9)\cdot e^{-2000\cdot r/159}\cdot (2\cdot 10^6\cdot r^2 - 954\cdot 10^3\cdot r + 75843)}{23121839365411671\cdot \pi}\cdot r^2$

#7: $1.376663340\cdot 10^{-8}\cdot r^2\cdot e^{-12.57861635\cdot r}\cdot (2\cdot 10^6\cdot r^2 - 9.54\cdot 10^5\cdot r + 7.5843\cdot 10^4)^2$

3p

#8: $\left| -r\cdot r\cdot e^{-r} \cdot \cos\left(\dfrac{\pi}{4}\right) \right|^2$

3p +/−

#9: $\left| -r\cdot r\cdot e^{-r} + i\cdot r\cdot e^{-r} \cdot (-r) \right|^2$

#10: $\left| -r\cdot r\cdot e^{-r} - i\cdot r\cdot e^{-r} \cdot (-r) \right|^2$

Identical : 180° rotated

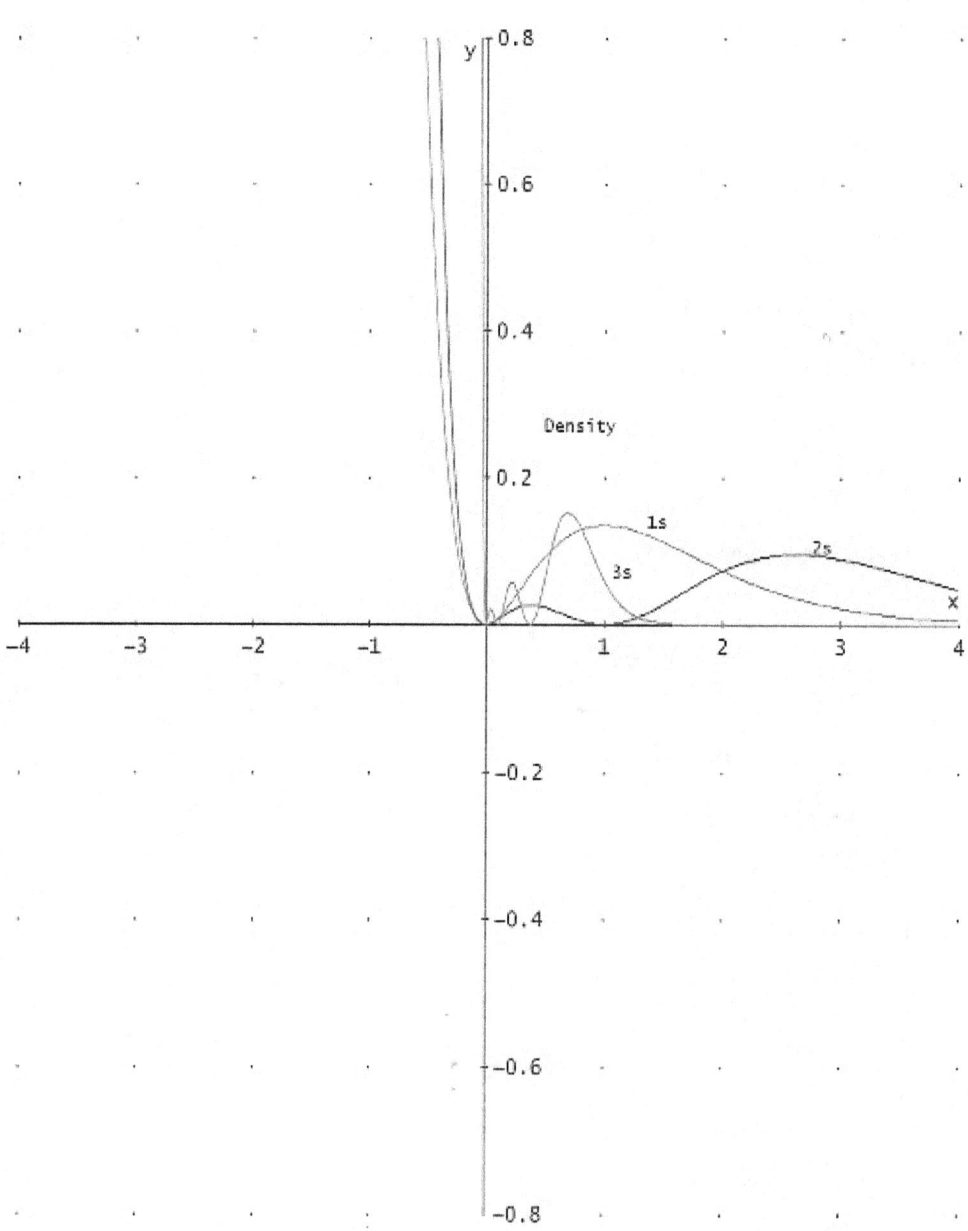

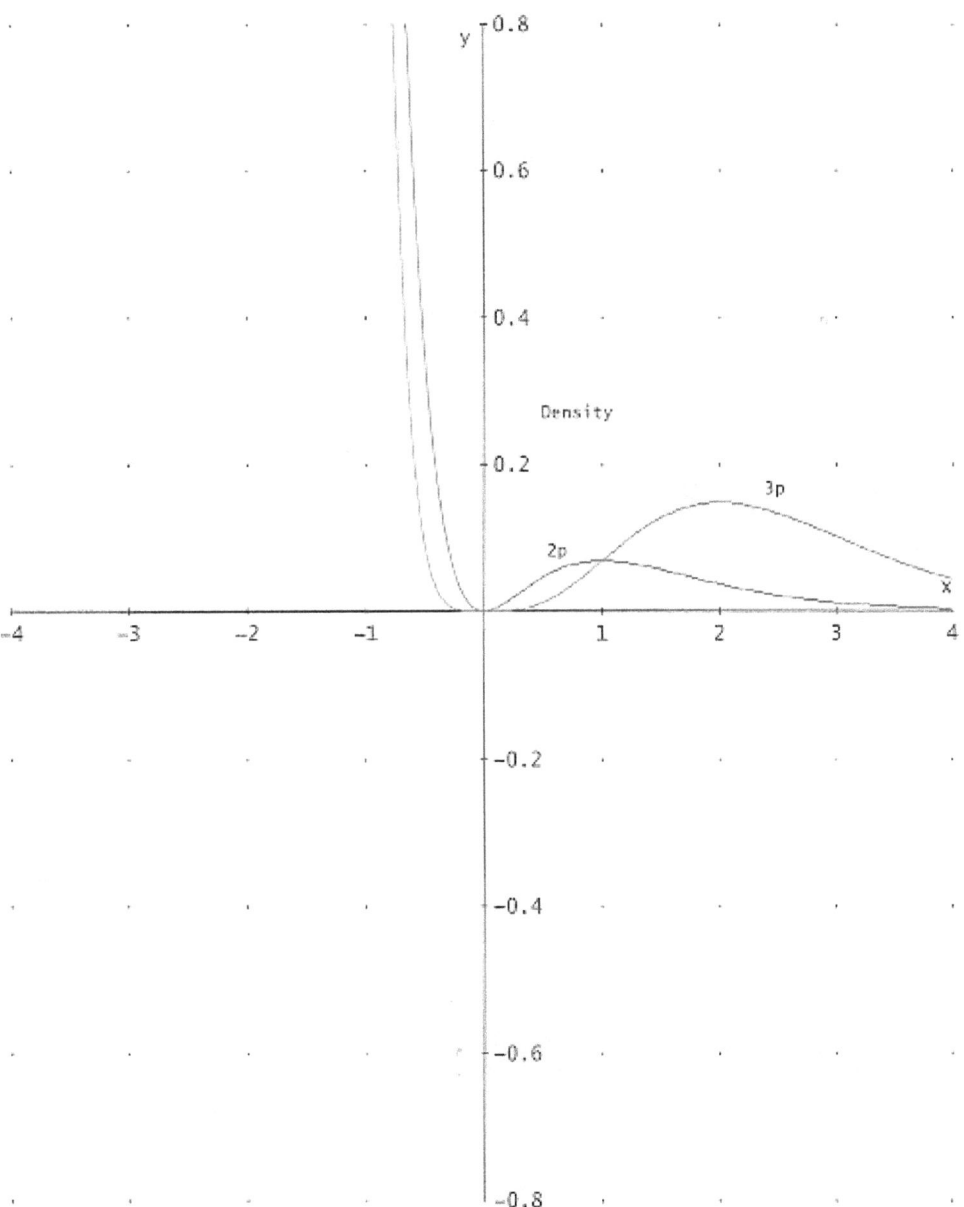

Note at the nucleus: The 3p has a large P(psi) = 0.

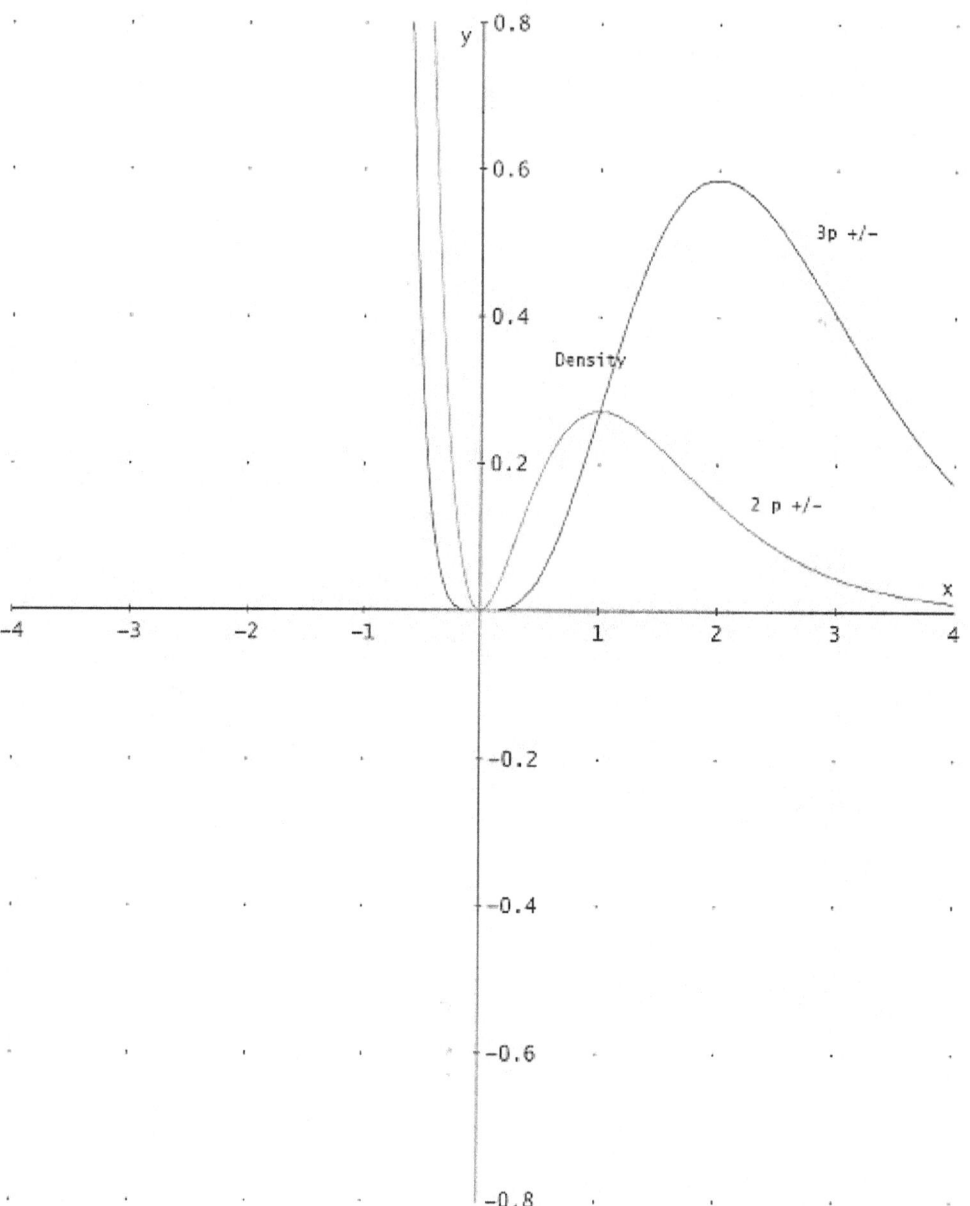

Note at the nucleus: The 3p +/- has a large P(psi)=0.

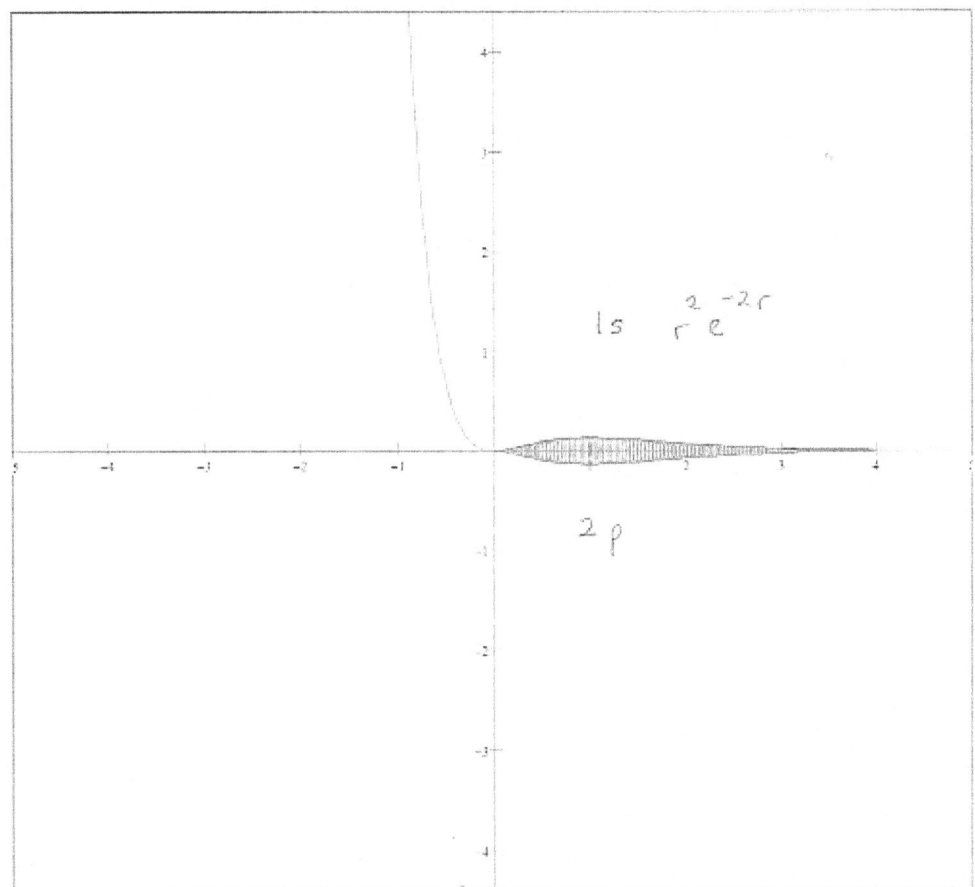

These solids of revolution try to conceptualize the actual shape of the shells and sub-shells.

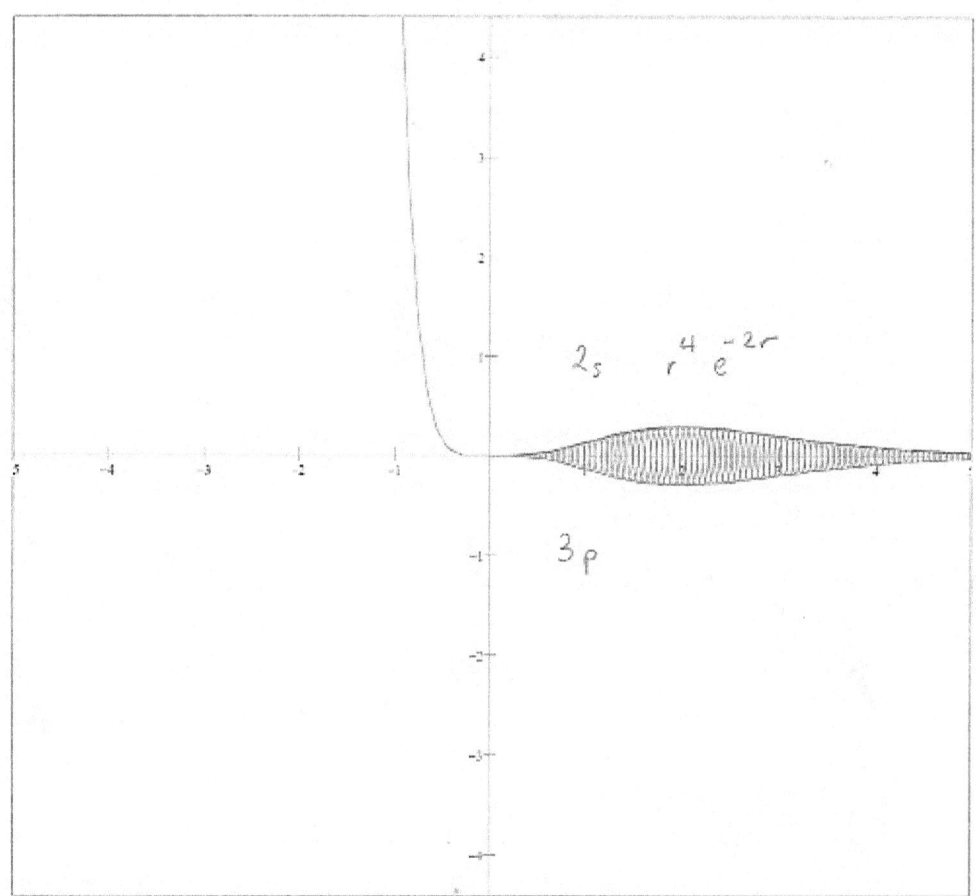

$2_s \qquad r^4 e^{-2r}$

3_p

Note the redundancy.
Note the enlargement of the shells.

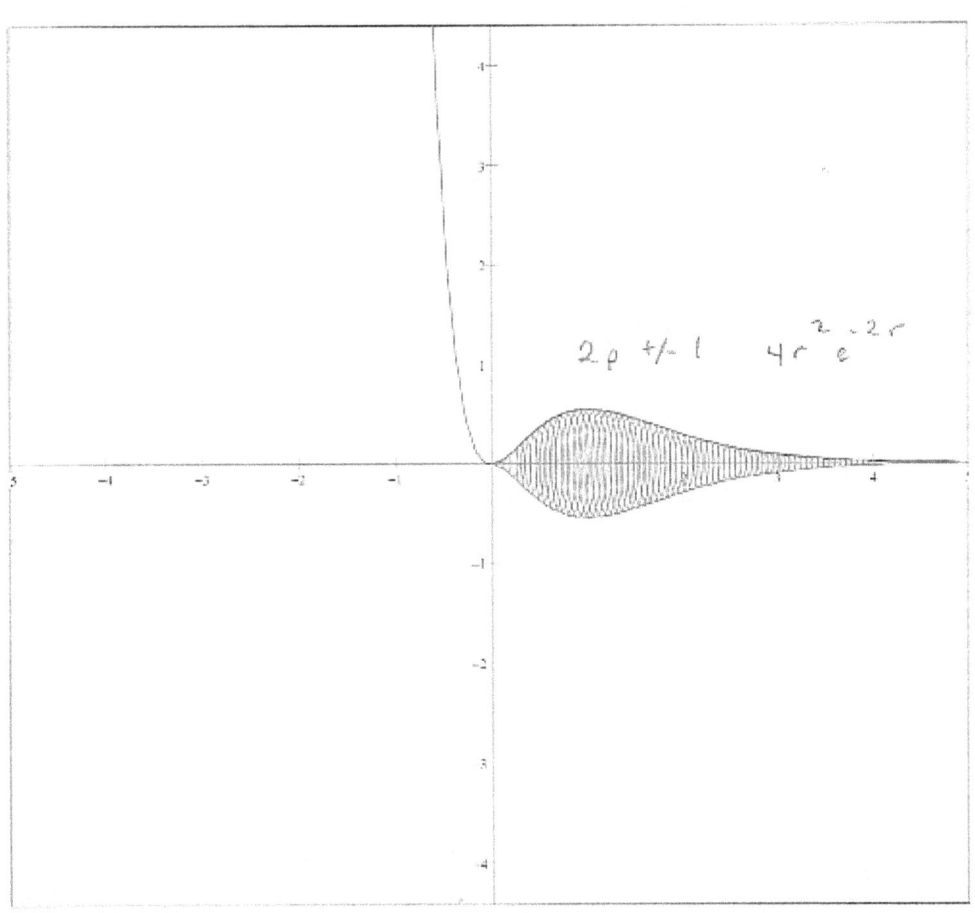

$2\rho +/- 1 \qquad 4r^2 e^{-2r}$

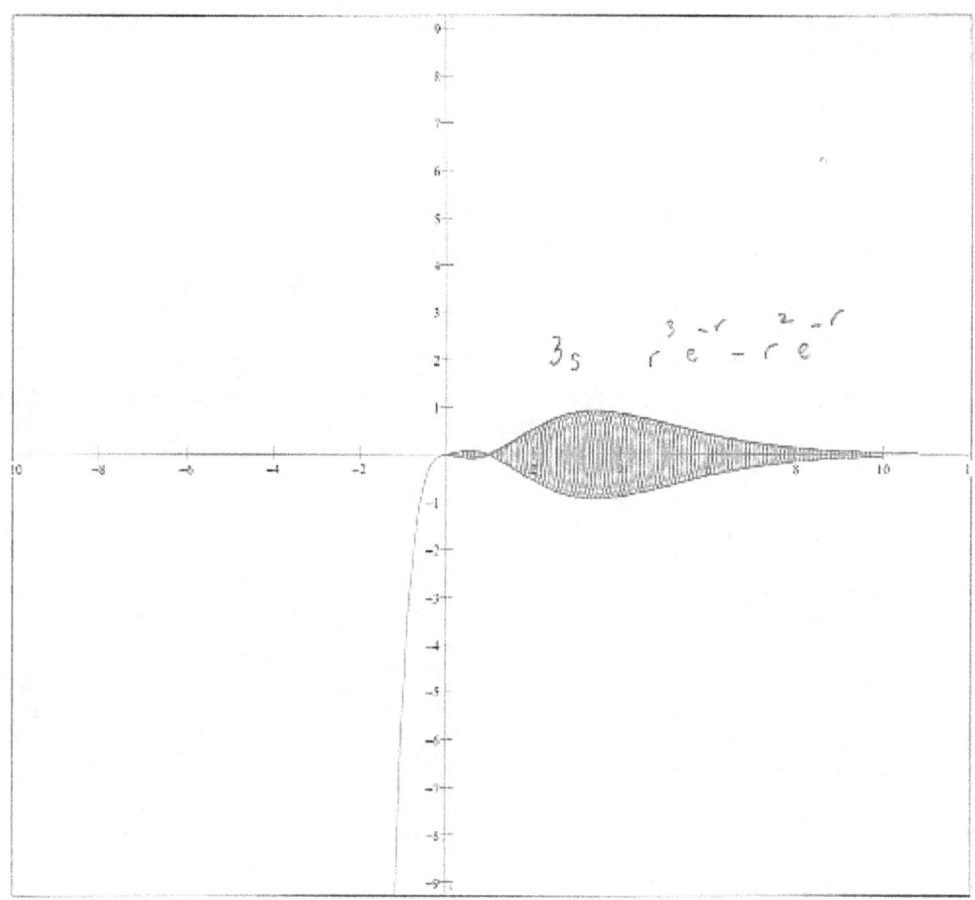

$$3_S \quad r^3 e^{-r} - r^2 e^{-r}$$

Note the double sphere geometry.

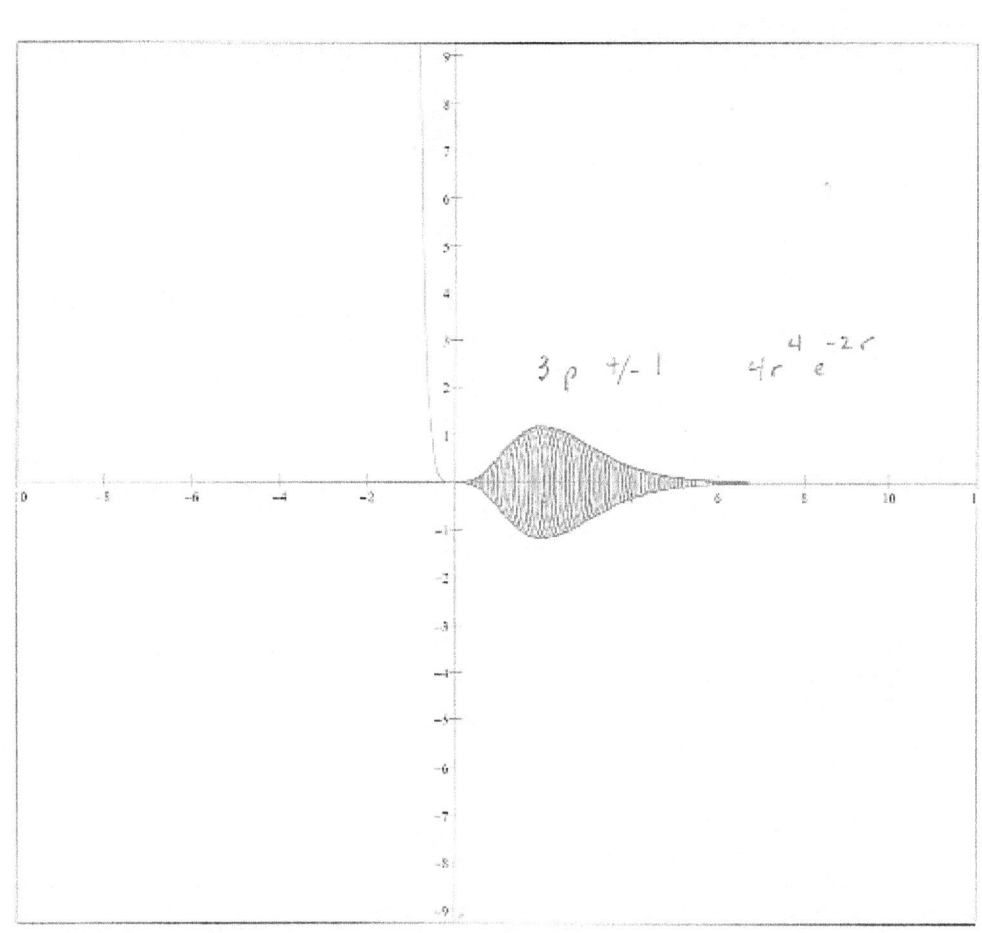

$3p +/- 1 \qquad 4r^4 e^{-2r}$

2 Orbit l

l "little script L" is the orbital Quantum Number

subshell: s=l=0
        p=l=1

$L=\sqrt{(l(l+1))}*h$ l=0,1,2,...,n-1)
Solution to Schrodinger equation with angular boundary conditions.
Discrete values of magnitude=quantized (allowed values).

#1: $a = \sqrt{(l\cdot(l + 1))}\cdot h$

a=orbital angular momentum
l=orbital quantum number
n=principle quantum number: determines the energy of a particular state.
h_=1.054*10^(-34) J*s

If l=0 then a=0. The electron cloud is spherically symmetric and has no
fundamental rotation axis. The particle is both wave and particle! It is
everywhere in a spherical volume.

Bohr model of the atom: For a Circle: L=angular momentum = m(e)vr.
L direction is 90° to the plane of circle (R. Hand Rule). Counter-
clockwise out of the plane. Clockwise is into the plane. The Bohr model
states that the p is quantized (h_). Even this qunatized model needs
modification. In the quantum world, there is no well-defined circle.

Rotational motion has an angular momentum character.

The Bohr Model of the atom: electron moves in a circle. r=radius. L=angular
momemtum relative to the center of the circle is L=m(e)vr.

m(e)vr= n*h_ restricted multiples of h_ (h bar). If L=0,this predicts that
the electrom passes through the nucleus-impossible!? Also the ground state
has an L=1?

The quantum-mechanical model has no circle. Classical model is concern
about the object being a particle only. The quantum model states that the
object is both particle and wave. There is no circle and no radius.

#2: $a = \sqrt{((-1)\cdot(-1 + 1))}\cdot h$

#3: $a = 0$

#4: $a = \sqrt{((-0.5)\cdot(-0.5 + 1))}\cdot h$

#5: $a = 0.5\cdot i\cdot h$

#6: $a = \sqrt{(0\cdot(0 + 1))}\cdot h$

#7: $a = 0$

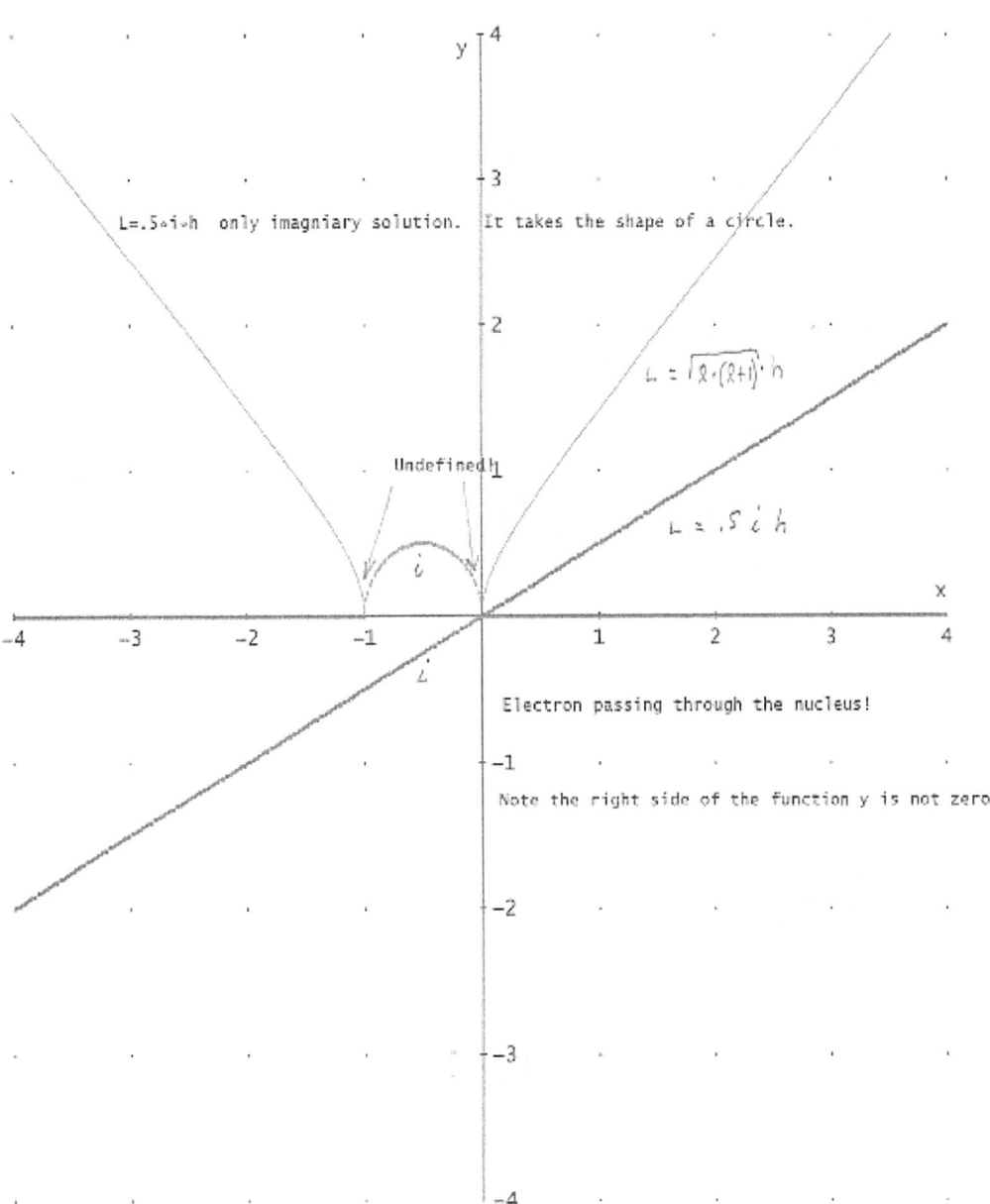

L=.5~i~h   only imagniary solution.   It takes the shape of a circle.

$L = \sqrt{\ell \cdot (\ell+1)} \cdot h$

Undefined!

$L = .5\, i\, h$

Electron passing through the nucleus!

Note the right side of the function y is not zero.

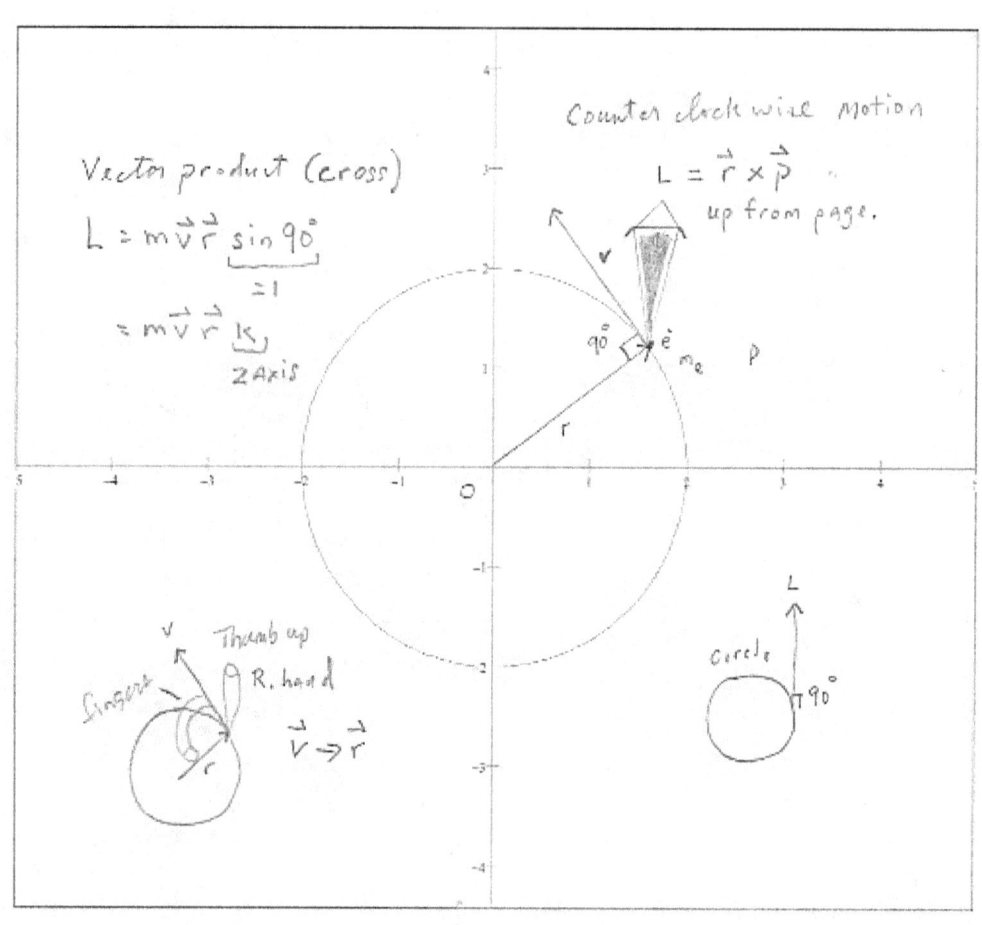

Vector product (cross)

$$L = m\vec{v}\,\vec{r}\;\underbrace{\sin 90°}_{=1}$$

$$= m\vec{v}\,\vec{r}\;\underbrace{k}_{z\,axis}$$

Counter clockwise motion

$$L = \vec{r} \times \vec{p}$$

up from page.

$90°$ $\dot{e}_{n_\theta}$ p

r

Thumb up
R. hand

v

fingers

r

$\vec{v} \to \vec{r}$

Circle

L

$90°$

3 Orbital Magnetic m(l)

```
m(l): Orbital Magnetic Quantum Number

The Bohr model: electron orbiting around the nucleus creates a current
loop (I). In Quantum, there is no circle, but there is an angular momentum
(L).  The electron being in rotational motion: L and μ (magnetic moment) is
present.

Current, Magnetic Field, Angular momentum, magnetic dipole moment
I        B              L                μ

I=Q/t
Q=charge
t=time

μ=IA
I=current
A=area (circle = πr^2

μ α L :  μ = [e/2*m(e)]*L
m(e)=mass of electron

Orbital motion in opposite directions cancel each other resulting in B=0.
The Spin magnetic moments cancel because of pairing of electrons in
opposite spin.
H atom has a magnetic moment of 9.27 *10^(-24) J/T because it has 1
upaired electron.
The total magnetic moment = Σ of orbital + spin moments.
The nucleus being proton and neutron also have moments but smaller
due to their larger mass.
```

Notes:

This is intrinsic, relativistic only. Not actual!

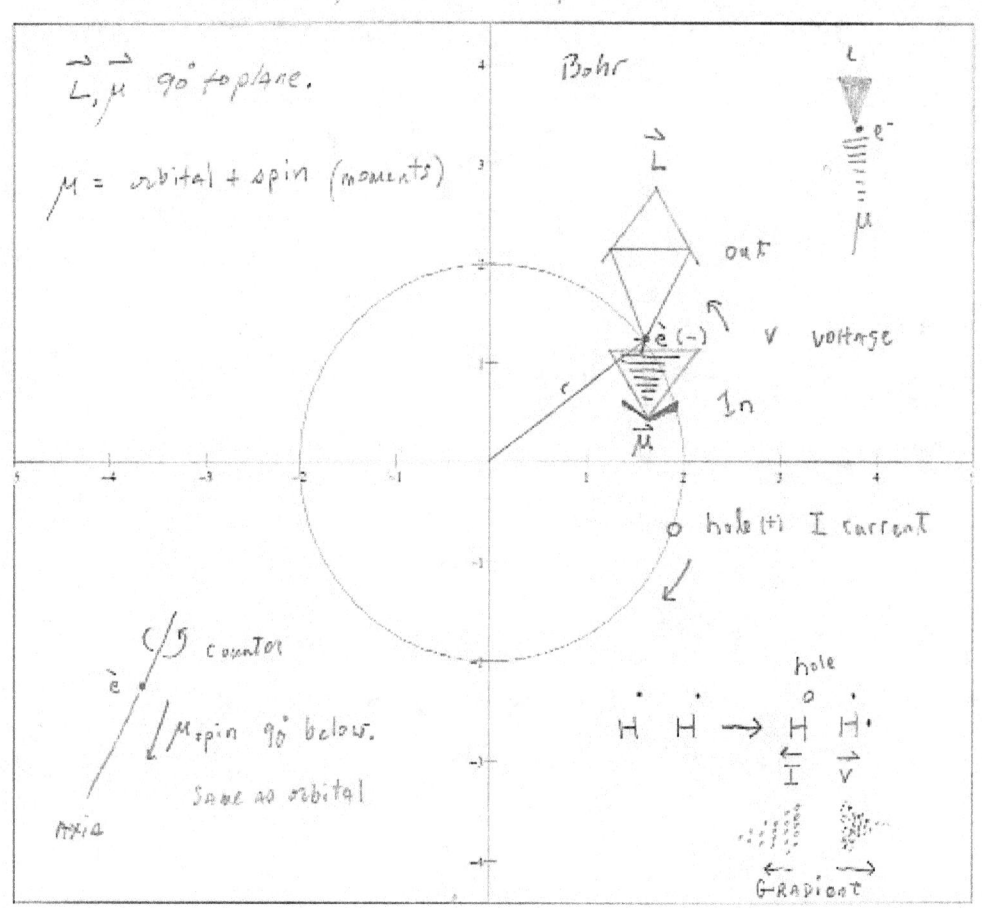

$\vec{L}, \vec{\mu}$ 90° to plane.

μ = orbital + spin (moments)

Bohr

out

$\vec{L}$

e (−)

V   voltage

In

$\vec{\mu}$

o hole (+) I current

e /s counter

μ spin 90° below.

Same as orbital

Axis

hole

o

H   H → H   H

I   V

Gradient

L,   L(z),   cosθ,
position vector   projected L on z axis   Angle between L and z axis

l
angular momentum

L(z) = m(l)* h_   allowed values of L(z)
                  quantized

```
l     m(l)              L(z)
0     0                 0
1     -1,0,1            -h_,0,h
2     -2,-1,0,1,2       -2h_,-h_,0,h_,2h_
```

cosθ = L(z)/|L| = m(l)/√(l*(l+1))

#1: $\cos(\theta) = \dfrac{a}{\sqrt{l \cdot (l + 1)}}$

#2: $\cos(\theta) = \dfrac{1}{\sqrt{l \cdot (l + 1)}}$

#3: $\text{SOLVE}\left(\cos(\theta) = \dfrac{1}{\sqrt{l \cdot (l + 1)}}, \ \theta\right)$

#4: $\theta = \text{ACOS}\left(\dfrac{1}{\sqrt{l \cdot (l + 1)}}\right) \lor \theta = \text{ASIN}\left(\dfrac{1}{\sqrt{l \cdot (l + 1)}}\right) + \dfrac{3 \cdot \pi}{2} \lor \theta = -$

$\text{ACOS}\left(\dfrac{1}{\sqrt{l \cdot (l + 1)}}\right)$

Quantized L orientations with an external magnetic field (B) is called
space quantization.

$$\sqrt{2}\,\hbar$$

L_z
$2\hbar$ θ $\sqrt{6}\,\hbar$ L

Space quantization

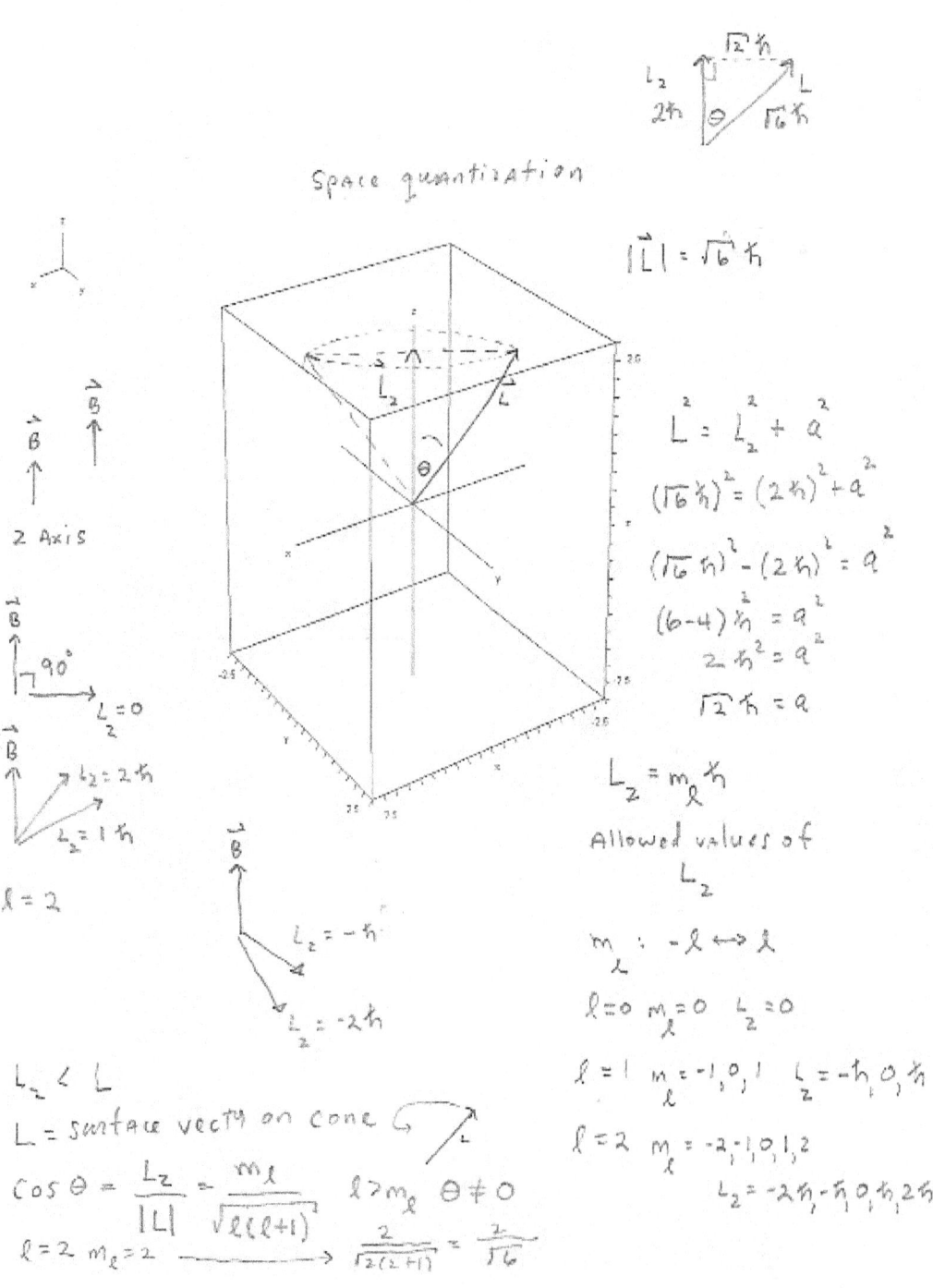

Z Axis

$\vec{B}$
$\vec{B}$
$\vec{B}$

$\vec{B}$
$90°$
$L_z = 0$

$\vec{B}$
$L_z = 2\hbar$
$L_z = 1\hbar$

$\ell = 2$

$\vec{B}$
$L_z = -\hbar$
$L_z = -2\hbar$

$|\vec{L}| = \sqrt{6}\,\hbar$

$$L^2 = L_z^2 + a^2$$

$$(\sqrt{6}\,\hbar)^2 = (2\hbar)^2 + a^2$$

$$(\sqrt{6}\,\hbar)^2 - (2\hbar)^2 = a^2$$

$$(6-4)\hbar^2 = a^2$$

$$2\hbar^2 = a^2$$

$$\sqrt{2}\,\hbar = a$$

$$L_z = m_\ell \hbar$$

Allowed values of
L_z

$m_\ell : -\ell \longleftrightarrow \ell$

$\ell = 0 \quad m_\ell = 0 \quad L_z = 0$

$\ell = 1 \quad m_\ell = -1, 0, 1 \quad L_z = -\hbar, 0, \hbar$

$\ell = 2 \quad m_\ell = -2, -1, 0, 1, 2$
$\qquad L_z = -2\hbar, -\hbar, 0, \hbar, 2\hbar$

$L_z < L$
L = surface vector on cone

$$\cos\theta = \frac{L_z}{|L|} = \frac{m_\ell}{\sqrt{\ell(\ell+1)}} \qquad \ell > m_\ell \quad \theta \neq 0$$

$\ell = 2 \quad m_\ell = 2 \longrightarrow \dfrac{2}{\sqrt{2(2+1)}} = \dfrac{2}{\sqrt{6}}$

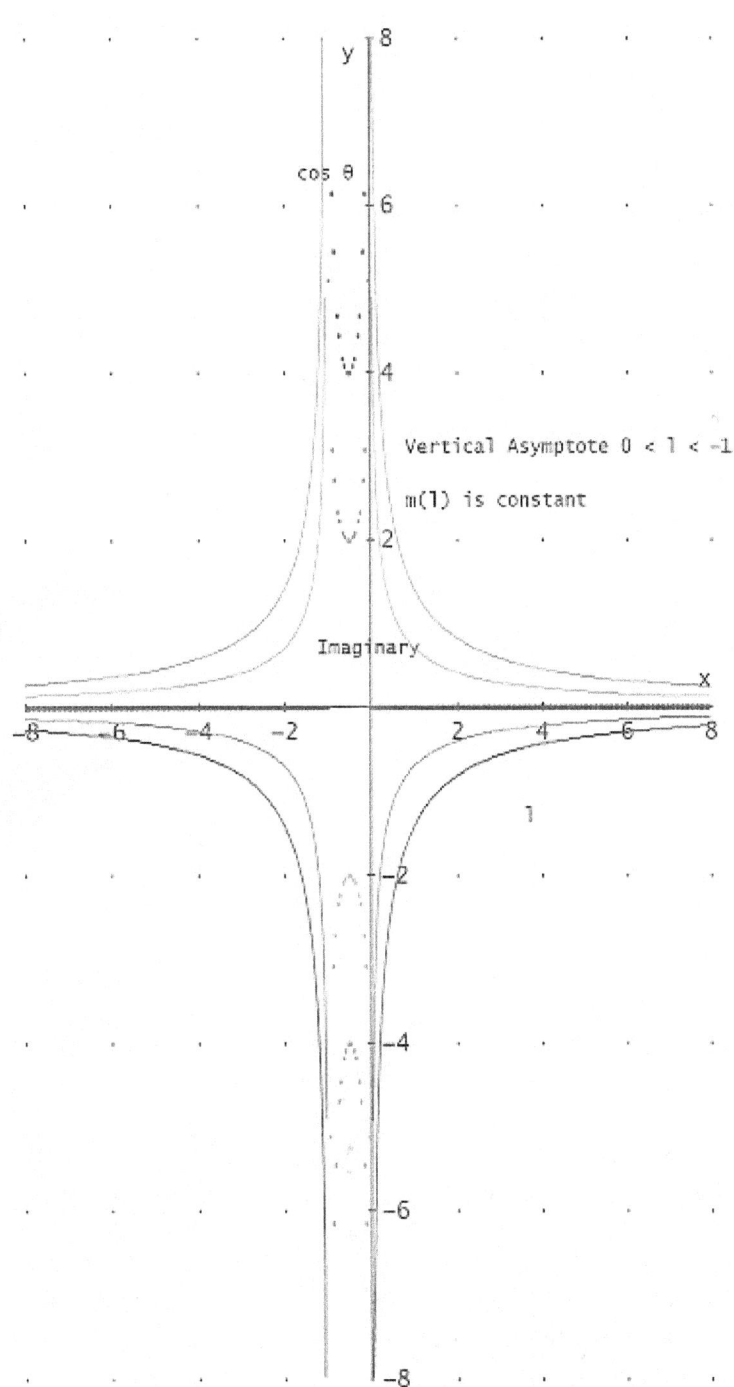

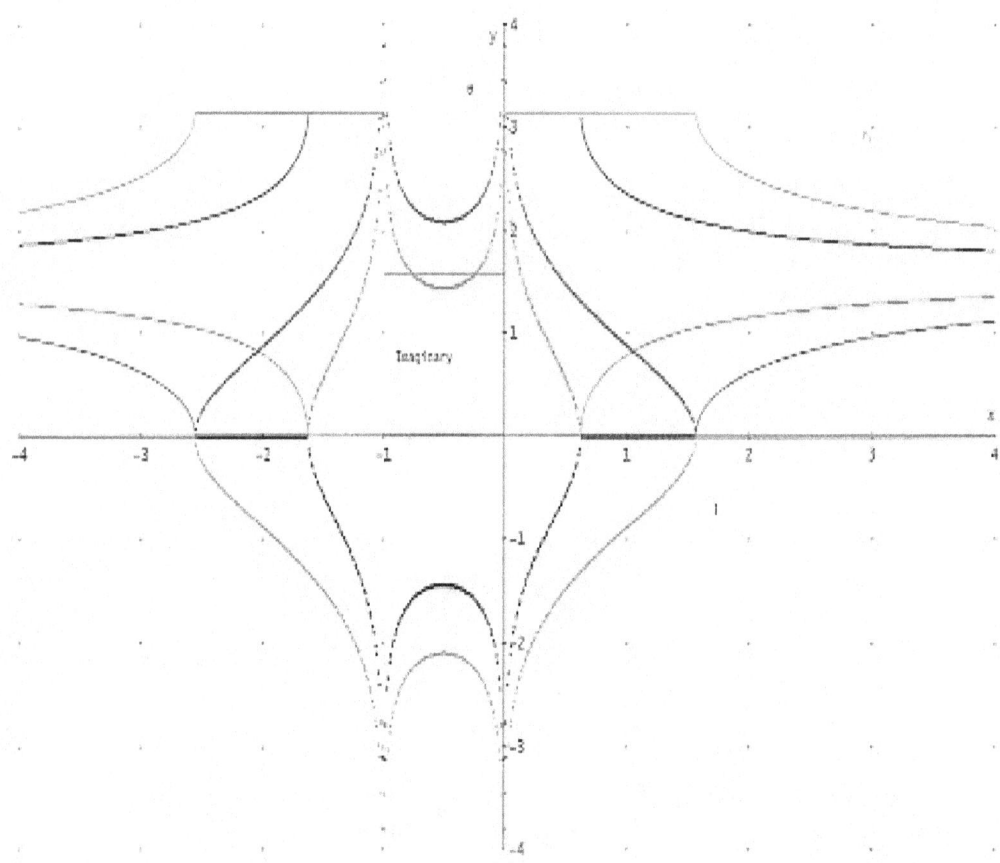

Cos theta moves toward the origin. Whereas theta moves away from the origin.
Cos theta is a much more appealing graph than theta.
But only the real values have significance.
The imaginary values are because of the radical.
  The existence of a external magnetic field in the environment of the atoms is either 0 or very small. Therefore the m(l) quantum number is of minor importance. Where it may become important is the unpaired electrons or ion state (the radical). In these states, m(l) plays a role in orientations of electron clouds for possible chemical reactions. 3D orientation, approximation of distance, lowering the energy barriers, and more all play a part in chemical reactions.

```
U= -u•B
U=potential energy
u=magnetic moment
B=magnetic field
The is additional energy for an atom in a magnetic field.
This U is added to the n energy.

Under a magnetic field effect, there are 3 directions of the
magnetic moment.  Thus creating 3 different energy levels between
2 subshells.

m(l)  hf(0)
1     h(f(0)+Δf)
0     hf(0)          B=0
-1    h(f(0)-Δf)
This is the Zeeman effect : 1 spectral line -> 3 spectral lines.
The excited atoms decay to the ground state with 3 different photons
with 3 different energies being emitted.
B creates a frequency shift of the emision (Δf).
```

Notes:

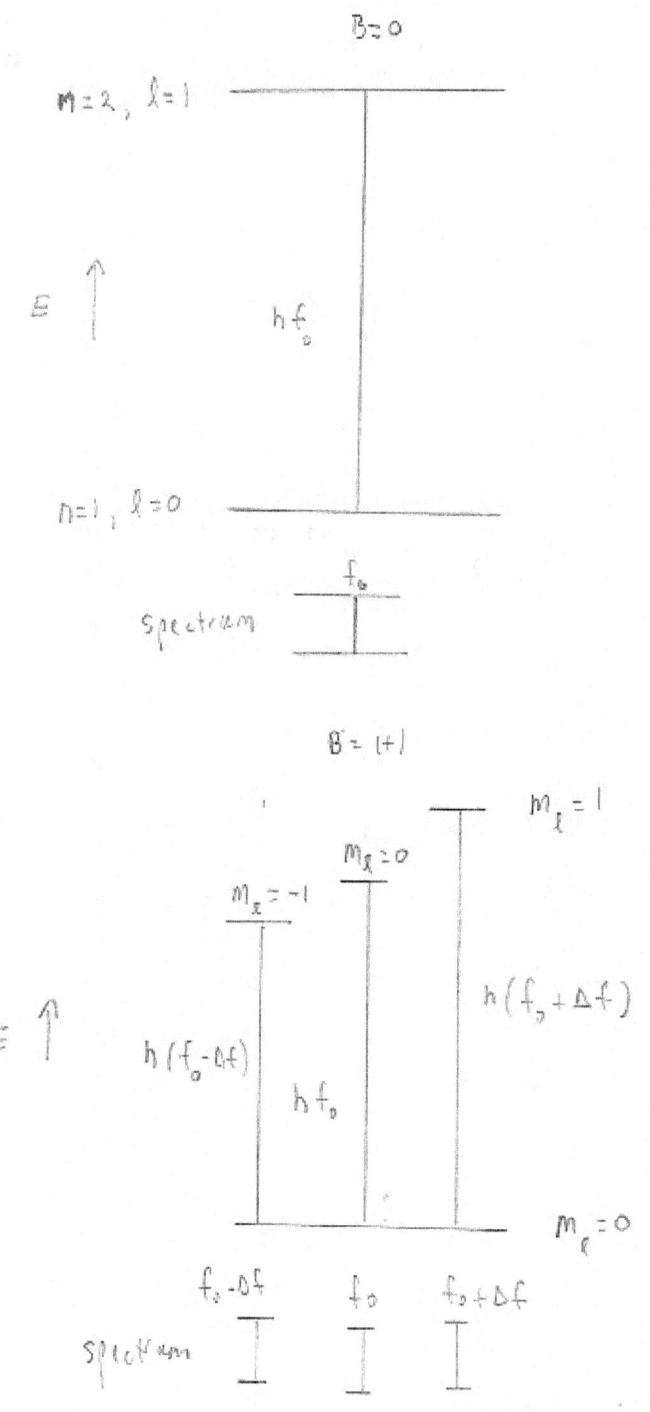

$B = 0$

$m = 2, \ell = 1$

$E \uparrow$

$h f_0$

$n = 1, \ell = 0$

f_0

spectrum

$B = (+)$

$m_\ell = 1$

$m_\ell = 0$

$m_\ell = -1$

$E \uparrow$

$h(f_0 - \Delta f)$

$h f_0$

$h(f_0 + \Delta f)$

$m_\ell = 0$

$f_0 - \Delta f$ f_0 $f_0 + \Delta f$

spectrum

4 Spin Magnetic m(s)

m(s):Spin Magnetic Quantum Number
Spin=rotation
Not derived from the Schrodinger equation
Emission spectrum demonstrates a "doublet" = double lines!
2 different energies is represented.
Spin up or Spin down are the 2 orientations.
This represents the relativistic properties of the electron: intrinsic angular momentum.
The electron is not truly spinning!
Spin s = +/- 1/2 and never changes.
+1/2 = spin up but deflected by the magnetic field B downward
-1/2 = spin down but deflected by B upward

$S = \sqrt{s(s+1)} * h\_ = \sqrt{3}/2\ h\_$ Magnitude $h\_$ = h bar

$\sqrt{[1/2(1/2+1)]} = \sqrt{(3/4)} = \sqrt{3}/2$

$S(z) = m(s)*h\_ = +/-h\_$

$\mu(spin) = -e/m(e) * S \rightarrow \mu(spin,z) = +/- eh\_/2m(e) = \mu(B).$

$L \propto h\_$ and $s \propto 1/2\ h\_$ angular momemtum is twice as great for spin than orbital.Substitute 1/2 h_ in above moment formula.

Bohr magneton $\mu(B)=9.27 \div 10^{(-24)}$ J/T

#1: $\mu = \dfrac{e \cdot h}{2 \cdot m}$

#2: $\mu = \dfrac{3}{325000000000000000000000}$

#3: $\mu = 9.230769230 \cdot 10^{-24}$

$e = 1.6 * 10^{(-19)}$
$m = 9.1 \div 10^{(31)}$
$h\_ = 1.05 \div 10^{(-34)}$

S can take a large angle to the z axis.

Using PT vector right triangle math yields: $1/2h\_$, $\sqrt{2}/2h\_$, and $\sqrt{3}/2h\_$.

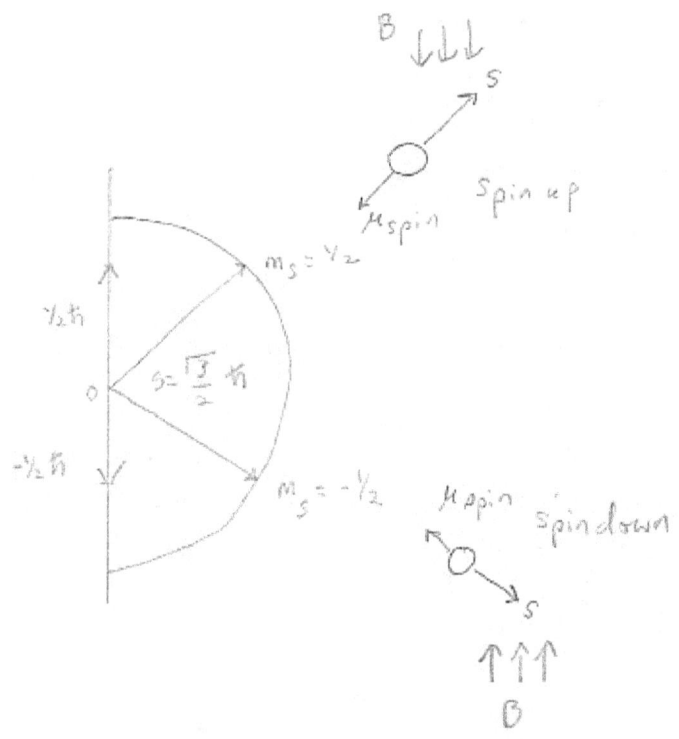

Can you do the vector addition? If you have the hypotenuse and 1 side, can you find the other side?

The magnetic field effect on the S is shown.

Pauli Exclusion Principle

@ orbit = 2 e (spin up and spin down). QN 4 dictates the final state of the atom. Opposite spins balance the 2 different states; almost equilibrating the energy levels of each state.

First shell n = 1 subshell s has 2 e. There is only 1 s. 1*2 = 2e. e configuration 1s(2) = [He].

Second shell n = 2 subshell s has 2 e and subshell p has 2 e. There is 1s and 3 p. 1*2+3*2 = 8e. e configuration 1s(2)+2s(2)2p(6) = 2+8 = 10e = [He]2s(2)2p(6) = [Ne].

Third shell n = 3 1s +3p+5d. 1*2+3*2+2*5=18e. e configuration 1s(2)2s(2)2p(6)3s(2)3p(6)=18e = [Ne]3s(2)3p(6) = [Ar].

Atomic number : Z = # p+ in atom's nucleus. Z = e at neutrality; eg. H(0) = 1 p+ + 1 e(-).

d orbitals

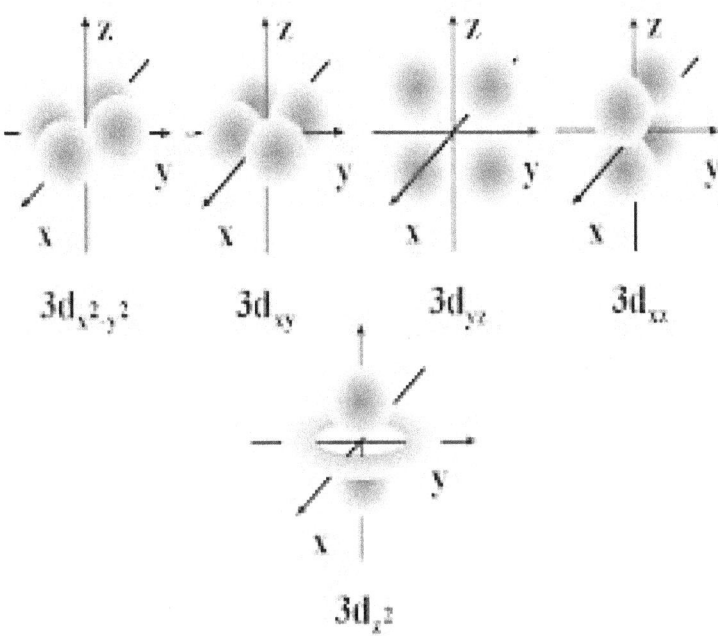

x^2 or y^2 : 2 double spheres on x and y axes or in the x and y planes.
xy, yz, and xz are 45 degrees off the axes as double spheres.
x^2 is a hybrid of all 3 axes: it covers the xy axes as a tubular circle and locates on the z axis also as double spheres.

Allowed QN States n=3 n,l,ml,ms u/d = up/down

n=1 (2)
100u/d

n=2 (8)
200u/d
211u/d 210u/d 21-1u/d

n=3 (18)
300u/d
311u/d 310u/d 31-1u/d
322u/d 321u/d 320u/d 32-1u/d 32-2u/d

Auflbau Principle : e fill the orbitals from the lowest to the highest energy levels first.

$$1s \rightarrow 2s \rightarrow 2p \rightarrow 3s \rightarrow 3p \rightarrow 3d$$

Hund's Rule : Degenerate orbits fill partially at first. Degenerate means same multiple energy levels. eg. p and d orbitals.

f orbitals

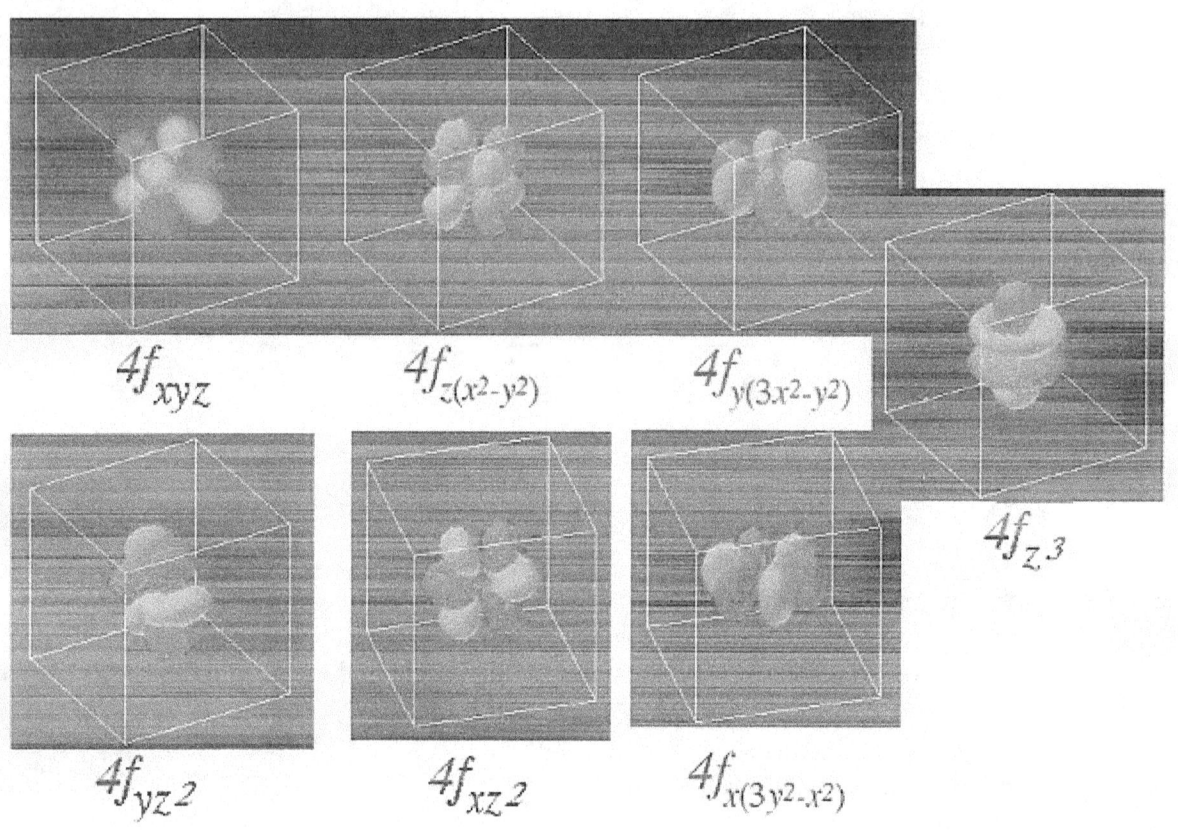

$$4f_{xyz} \qquad 4f_{z(x^2-y^2)} \qquad 4f_{y(3x^2-y^2)} \qquad 4f_{z^3}$$

$$4f_{yz^2} \qquad 4f_{xz^2} \qquad 4f_{x(3y^2-x^2)}$$

Notes:

Summary of important orbitals

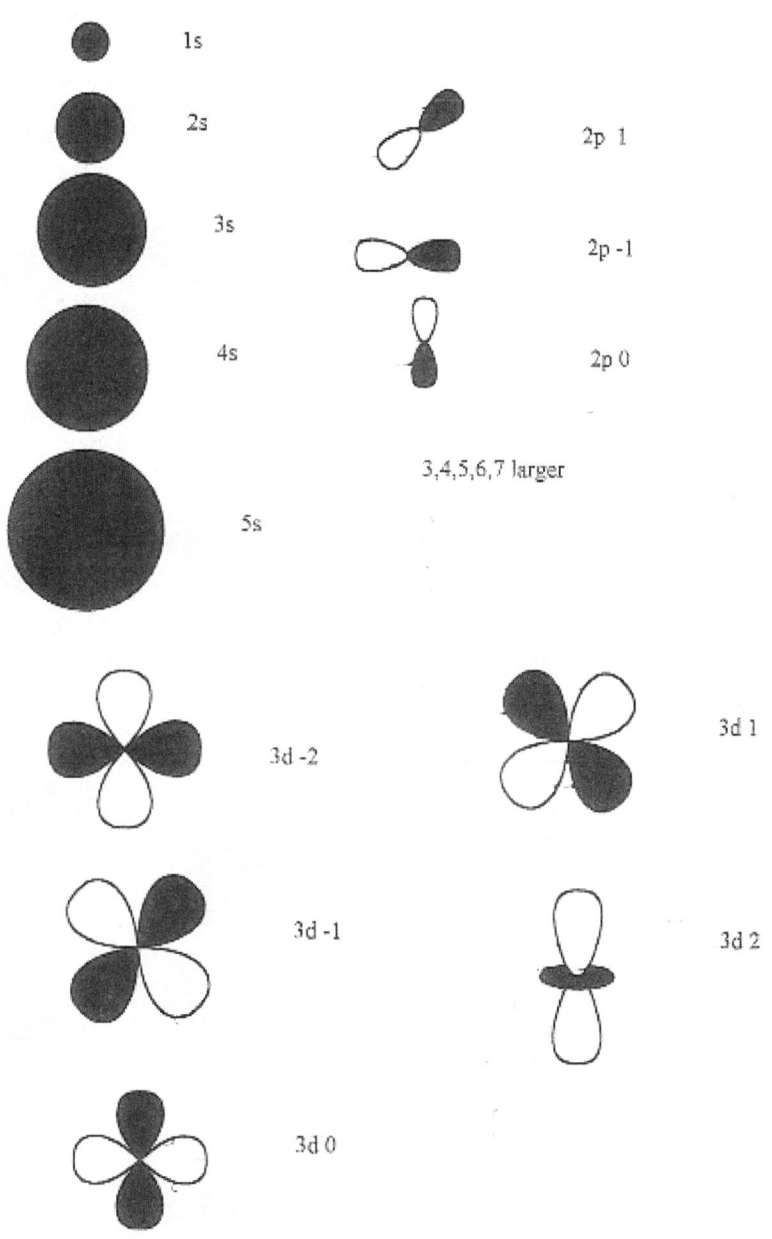

Electron Configurations:

Z	Symbol	Ground-State Configuration	Ionization Energy (eV)	n
1 H	1s1 1st Row		13.595	1
2 He	1s2		24.581	1
3 Li	[He]2s1 2nd Row	Aufbau principle	5.39	2
4 Be	[He]2s2		9.32	2
5 B	[He]2s22p1		8.296	2
6 C	[He]2s22p2		11.256	2
7 N	[He]2s22p3		14.545	2
8 O	[He]2s22p4		13.614	2
9 F	[He]2s22p5		17.418	2
10 Ne	[He]2s22p6		21.559	2
11 Na	[Ne]3s1 3rd Row		5.138	3
12 Mg	[Ne]3s2		7.644	3
13 Al	[Ne]3s23p1		5.984	3
14 Si	[Ne]3s23p2		8.149	3
15 P	[Ne]3s23p3		10.484	3
16 S	[Ne]3s23p4		10.357	3
17 Cl	[Ne]3s23p5		13.01	3
18 Ar	[Ne]3s23p6		15.755	3
19 K	[Ar]4s1 4th Row		4.339	3
20 Ca	[Ar]4s2		6.111	3
21 Sc	[Ar]3d14s2	3d<4s	6.54	3
22 Ti	[Ar]3d24s2		6.83	3
23 V	[Ar]3d34s2		6.74	3
24 Cr	[Ar]3d54s1		6.76	3
25 Mn	[Ar]3d54s2		7.432	3
26 Fe	[Ar]3d64s2		7.87	3
27 Co	[Ar]3d74s2		7.86	3
28 Ni	[Ar]3d84s2		7.633	3
29 Cu	[Ar]3d104s1		7.724	4
30 Zn	[Ar]3d104s2		9.391	4
31 Ga	[Ar]3d104s24p1	3d<4s<4p	6	4
32 Ge	[Ar]3d104s24p2		7.88	4
33 As	[Ar]3d104s24p3		9.81	4
34 Se	[Ar]3d104s24p4		9.75	4
35 Br	[Ar]3d104s24p5		11.84	4
36 Kr	[Ar]3d104s24p6		13.996	4
37 Rb	[Kr]5s1 5th Row		4.176	4
Cs	6th Row			
Fr	7th Row			
La	Lanthanide Series f orbit			
Ac	Actinide Series f orbit			

Energy   Quantum Level
increasing   1s   Lowest
              2s
              2p
              3s

Sheet1

```
3p
4s
3d
4p
5s
4d
5p
6s
5d              -> 4f Lanthanide series        d<f
5p
7s      highest
6d              -> 5f Actinide series          f<d
```

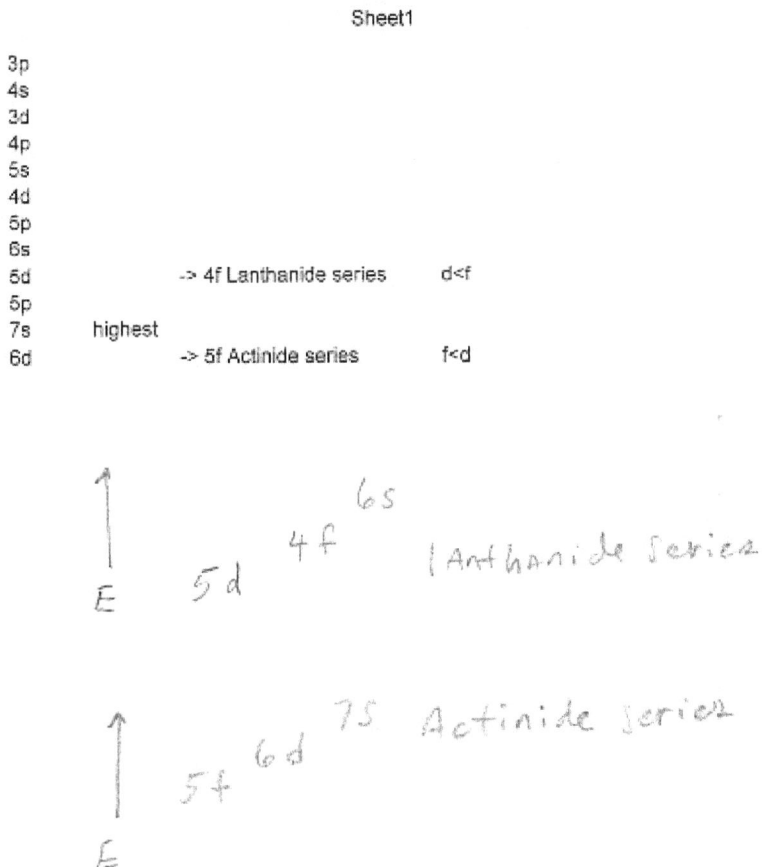

Notes:

Sheet1

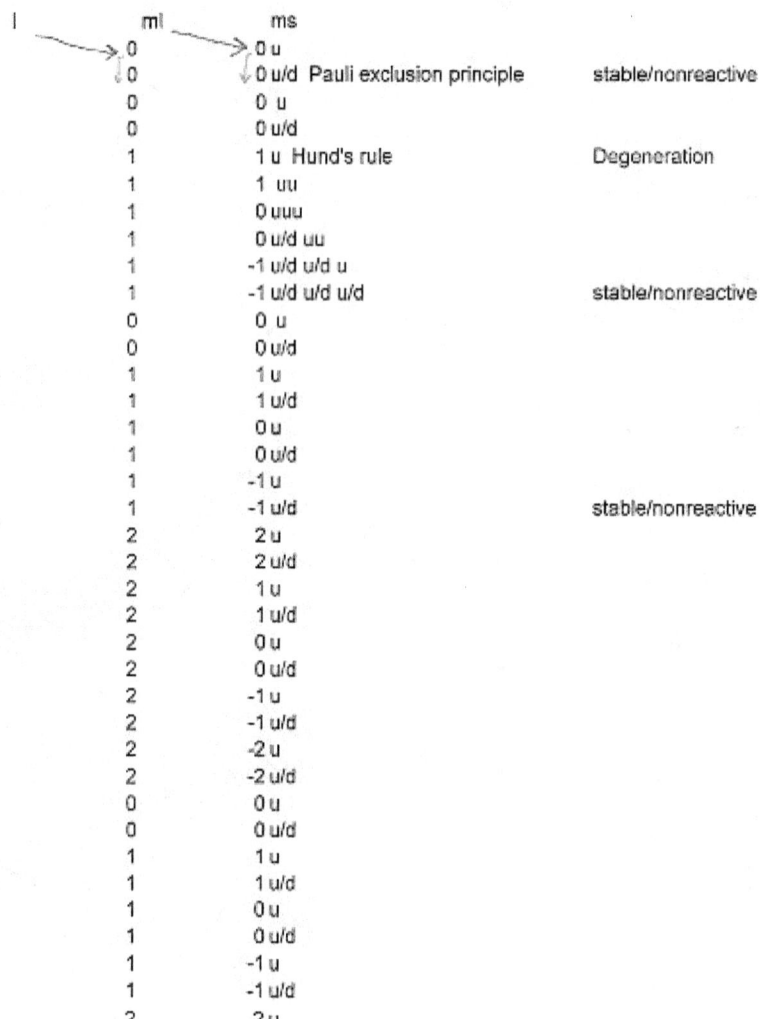

l	ml	ms		
0	0	u		
0	0	u/d	Pauli exclusion principle	stable/nonreactive
0	0	u		
0	0	u/d		
1	1	u	Hund's rule	Degeneration
1	1	uu		
1	0	uuu		
1	0	u/d uu		
1	-1	u/d u/d u		
1	-1	u/d u/d u/d		stable/nonreactive
0	0	u		
0	0	u/d		
1	1	u		
1	1	u/d		
1	0	u		
1	0	u/d		
1	-1	u		
1	-1	u/d		stable/nonreactive
2	2	u		
2	2	u/d		
2	1	u		
2	1	u/d		
2	0	u		
2	0	u/d		
2	-1	u		
2	-1	u/d		
2	-2	u		
2	-2	u/d		
0	0	u		
0	0	u/d		
1	1	u		
1	1	u/d		
1	0	u		
1	0	u/d		
1	-1	u		
1	-1	u/d		
2	2	u		

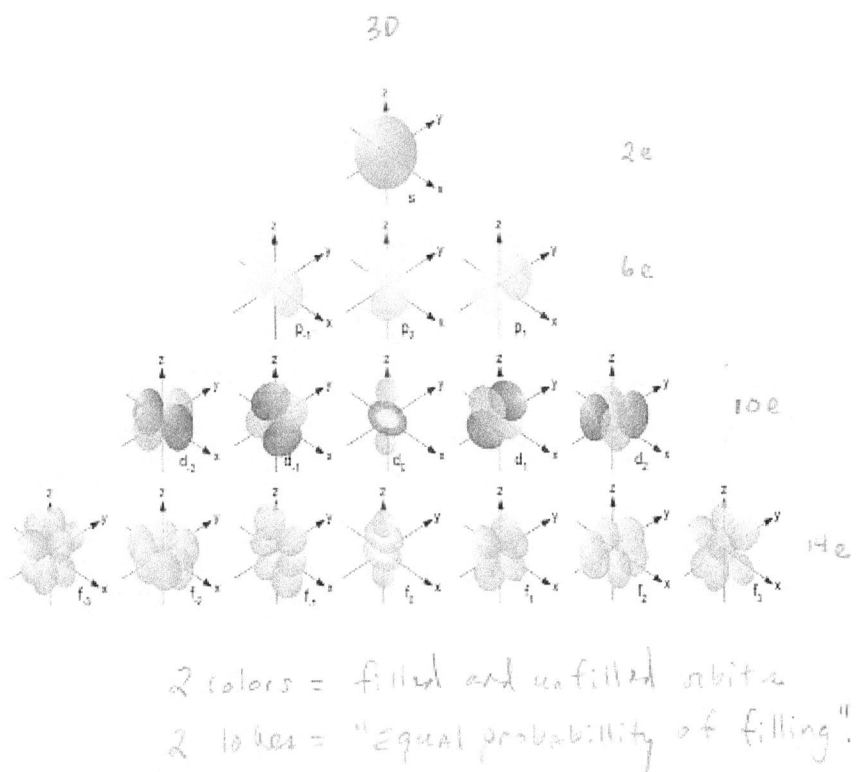

3D

2e

6e

10e

14e

2 colors = filled and unfilled orbitals

2 lobes = "Equal probability of filling".

As the atom fills from s->p->d->f, the orbitals are adding and superimposing. Th ultimate shape at the end would be generally spherical.

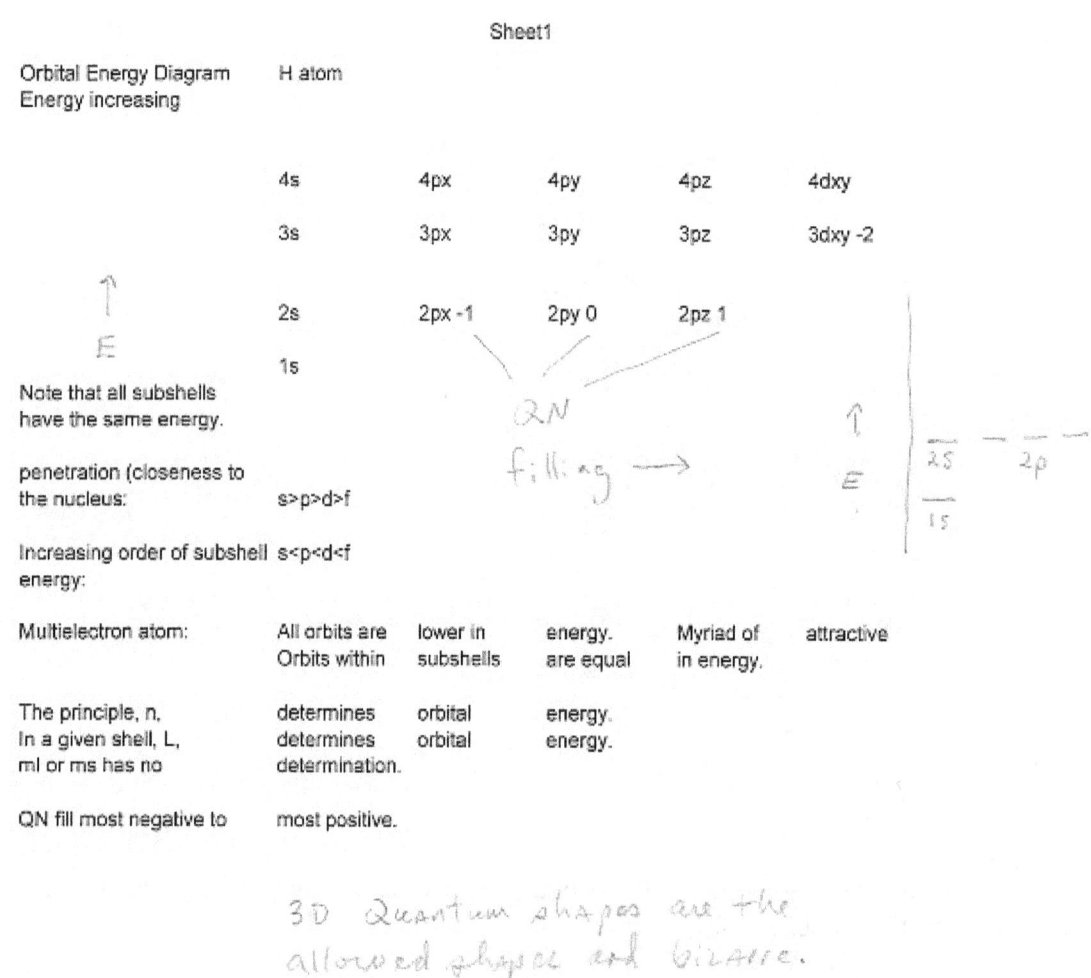

Orbital Energy Diagram   H atom
Energy increasing

	4s	4px	4py	4pz	4dxy
	3s	3px	3py	3pz	3dxy -2
	2s	2px -1	2py 0	2pz 1	
	1s				

Note that all subshells
have the same energy.

penetration (closeness to
the nucleus: s>p>d>f

Increasing order of subshell s<p<d<f
energy:

Multielectron atom:	All orbits are Orbits within	lower in subshells	energy. are equal	Myriad of in energy.	attractive
The principle, n, In a given shell, L, ml or ms has no	determines determines determination.	orbital orbital	energy. energy.		

QN fill most negative to   most positive.

The Orbital Energy Diagram is an easy way of portraying the e in each orbit as it relates to the energy level.

This table was created by a spreadsheet. It reads left to right.

4dyz   4dz^2   4dxy   4d(x^2-y^2)   4fz(x^2-y^2) -3   4fy(3x^2-y^2) -2

3dyz -1   3z^2 0   3dxy 1   3d(x^2-y^2) 2

and repulsive forces.

4fyz^2 -1   4fz^3 0   4fxz^2 1   4fx(3y^2-x^2) 2   4fxyz 3

These are vertical sections that should be placed side by side to be read left to right.

```
Summary Electron Configuration:
2 types of notation:
1) spdf notation:  1s^2 2s^2 2px^1 2py^1 2pz^1
   subshell                axis        # e
2) Orbital Diagram:  1s 2s 2p
                     ud ud u u u spin QN
                     arrows
                           parallel spin
                           opposing spins are paired.

There are exceptions to the filling process which is driven by
overlapping energy levels and the atom seeking the lowest energy state.
This was established by experiments using spectroscopy and magnetic
studies.
```

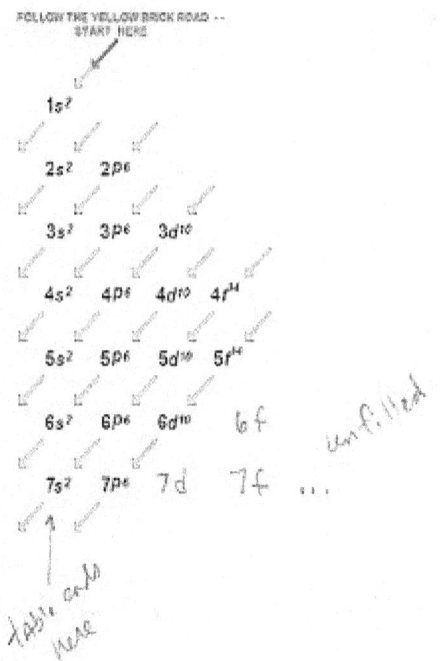

Notes:

Sheet1

Principle shell : n		1	2	3	4 ... n
Subshell : l		0 0 1	0 1 2	0 1 2 3	
Subshell Designation	s	s p	s p d	s p d f	
Orbitals in Subshell		1 1 3	1 3 5	1 3 5 7	
Subshell Capacity		2 2 6	2 6 10	2 6 10 14	
Principle shell Capacity		2	8	18	32 ... $2n^2$

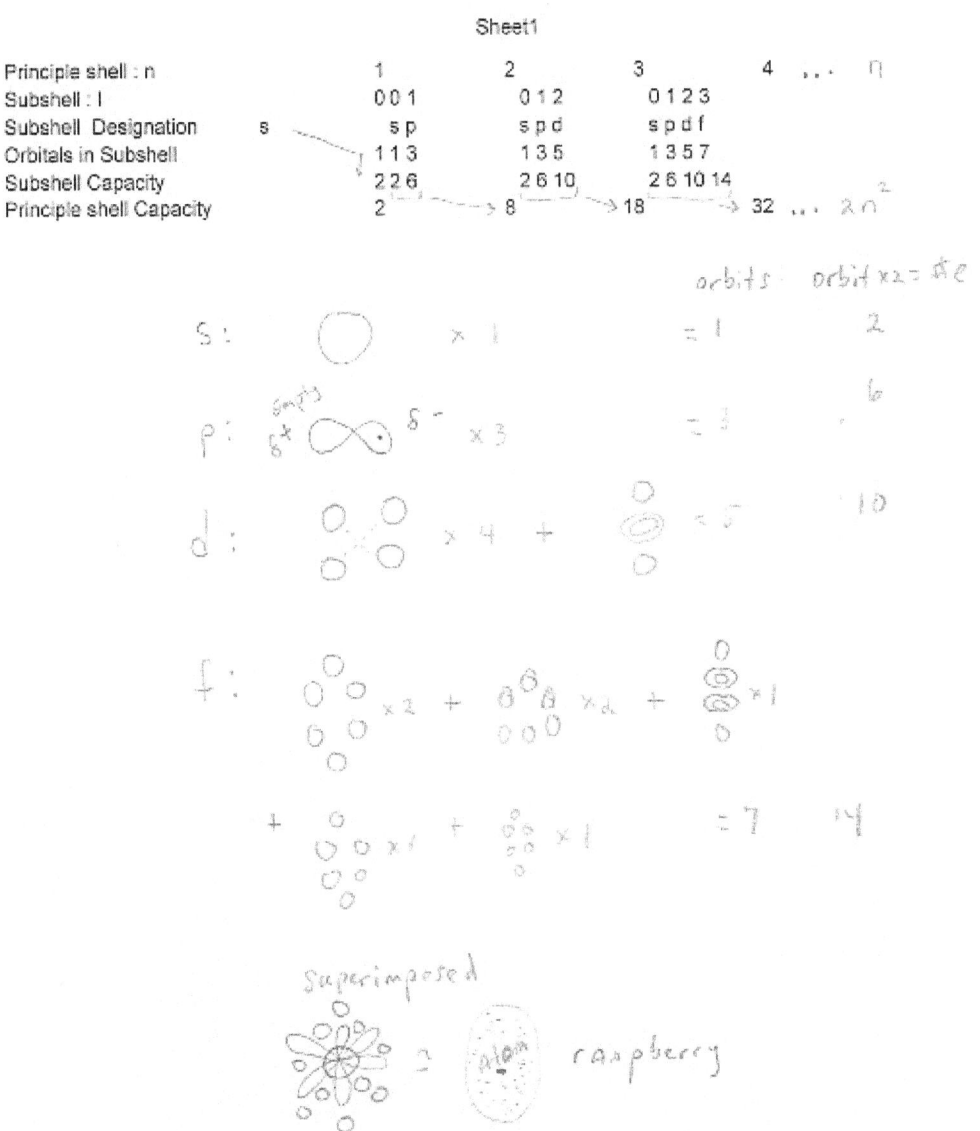

s-block is A.

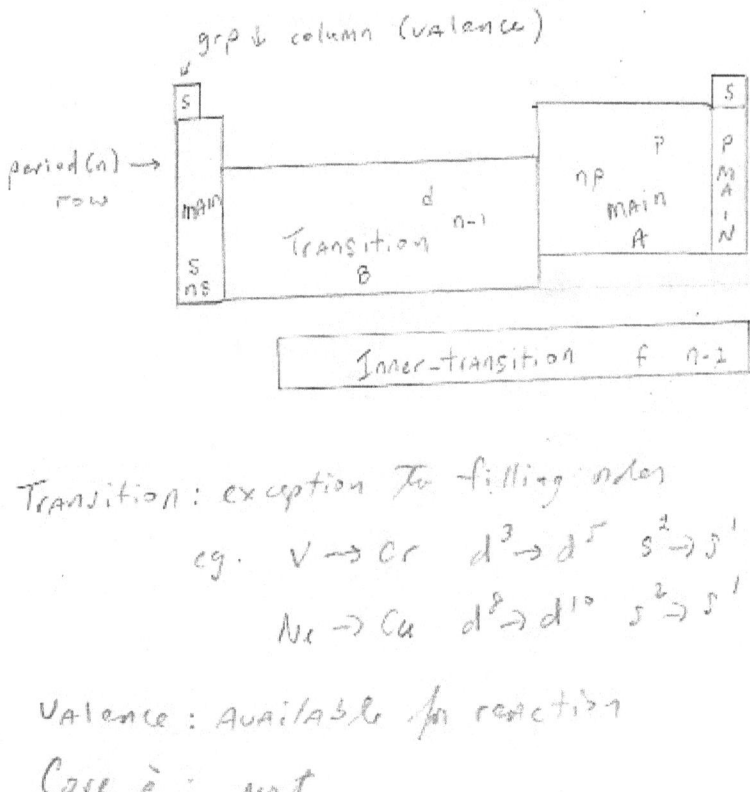

Transition: exception to filling order

eg. V → Cr $d^3 → d^5$ $s^2 → s^1$

Ni → Cu $d^8 → d^{10}$ $s^2 → s^1$

Valence: available for reaction

Core e⁻: not

More nonmetallic, more negative EA, increasing I1 →   and up the right side.

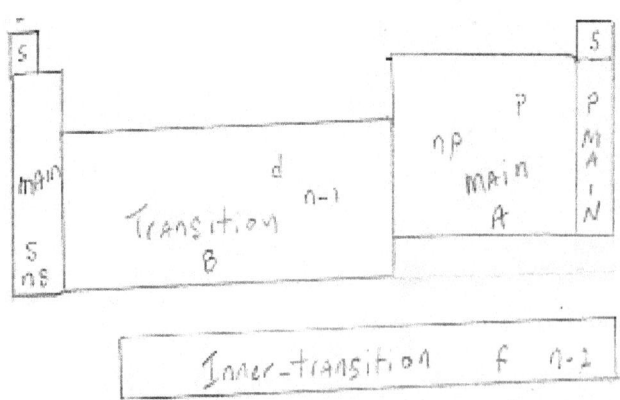

<--- Increasing atomic radius, more metallic and down the left side.

EA : electron affinity is the energy change with the addition of an electron.
I1 : ionization energy is the energy required to remove an electron.

In the n=4 period, the d starts filling at 3d (n-1).
In the n=6 period, the f starts filling at 4f (n-2).

Magnetic Spin and Magnetic Properties:

The e is orbiting around the nucleus in a circular motion. The circle is a looped path. The current is in the direction of e flow and a magnetic field is perpendicular, Right Hand Rule, to this flow.
Pair of e have opposite spins and cancel each other.
Unpaired e has a magnetic field.

Diamagnetism = weak repulsion of paired e in an external magnetic field.
Paramagnetism = strong attraction in an external magnetic field.
Ferromagnetism = very strong e. Fe (iron).

Atomic Radii :

Covalent: ½ distance between diatomic molecule.
Metallic : ½   "   '   adjacent atoms in solid metal.
Ionic : ½ diameter of each ion added together.
        Cation : smaller
        Anion : larger
Greater the nuclear charge; the smaller the specie.

Ionization Energy :

Energy required to remove 1 e.
Valence e are held weakly compared to the core e.

Electron Affinity :

The energy changed that occurs with the addition of an e. Energy is released in the process, exothermic. Therefore the negative quantity.

125

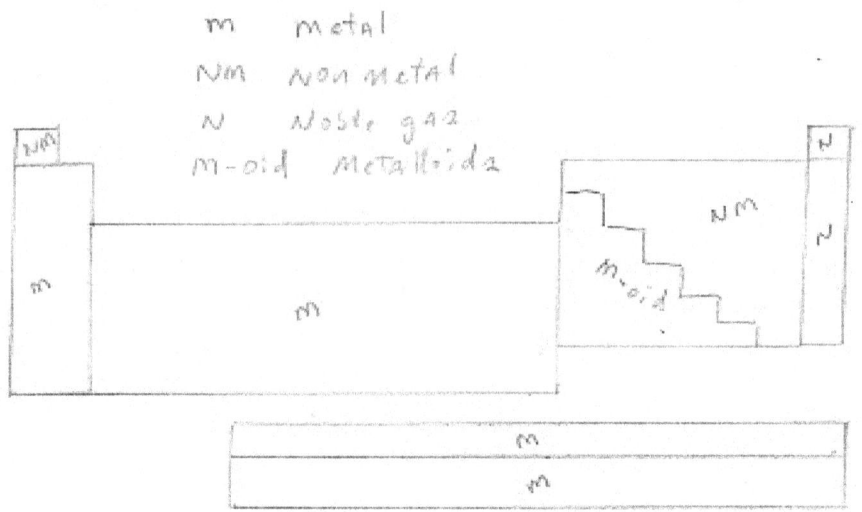

Flame Colors :

Group 1A : metals
Group 2A : heavier metals

Both have lowest I1.

Heat → valence e elevated to higher energy level (subshell) drops back to a lower energy level and emits energy as photons (radiation).

Halogens as Oxidizing Agents :

More (-)EA : extracts e from other atom.

s-Block Metals as Reducing Agents :

Low I1 energy : gives up e to the other atom. The donating atom is oxidized. But is considered a reducing agent.

Acidic, Basic, and Amphoteric Oxides :

Nonmetallic oxide → acid
Metallic oxide → base
Metalloid oxide can be amphoteric = act as acid or base. eg. Al Group 3A.

Gilbert Newton Lewis

Octet Rule :

A chemical bond joins two atoms together to form a molecule.

The valence e are the ones involved in this process.

The rule implies that the atoms in the combination want to reach 8 e in their outer shell. This is the noble gas state which is very stable and thermodynamically acceptable.

There are 2 main types of bonds:

1. Ionic : Atoms on opposite sides of the periodic table.
2. Covalent: Atoms in the center of the table.

eg. of ionic: NaCl : Na loses 1 e → Neon Cl gains 1 e → Ar Ions are created (radicals). Electrostatic attraction is in play. A salt is created (crystal). This in mainly an inorganic reaction.

The proper term is ionic interaction. Electrons are transferred, not shared. The charges attract each other.

eg. of covalent: C-C : It would take a lot of energy to transfer 4 e. So instead, C will share it's e.

The ground state is lowest energy level for the atom. For C: 2s 2p = 4 e for filling (8 - 4 = 4): C^0.

Atomic orbitals give birth to Molecular orbitals

When 2 atoms come together, they seek the lowest potential-energy state and optimal internuclear distance.

Lewis Symbol: C with 4 dots. The symbol of the element represents the core of the atom and the dots the valence electrons.

Lewis Structure: H : H or H – H. The bond is represented as a straight line. It contains 2 electrons or a bonding pair.
Lone pairs are not in bonds.

EN, Electronegativity, creates a polarity in the atoms in a molecule.
There are 2 types: polar and non-polar.
The atom with the higher EN is more negative.
Symbols to designate are delta +/- or an arrow with a + tail. The delta means partial. The arrow with the + tail pointing to the head, which is (-), depicts polarity.

Formal charge is the electron difference between bonded and non-bonded electrons in an atom in a molecule.

FC = # valence e - # lone pair - ½ # bonded e C (2 LP)s = S = S(2 LPs).

C : 4 – 4 - ½ 4 = -2

S : 6 – 0 - ½ 8 = +2

S : 6 – 4 - ½ 4 = 0

If there are FC on atoms, this is called a coordinate covalent bond.

Resonance is delocalized bonding. Molecules are identical in structure but can be hybrid in e location. From an e perspective, the e vibrates between atoms and an averaging takes place.
A double-headed arrow or a dotted line symbolizes this condition.

Radicals: Odd # valence e : very reactive
Deficient # : Be, B, or Al.
Expanded # : P, S

Lewis Symbol:

Group 1A Period 2 : Li

Li ·   1e

———————→   Valence

Group 8A Period 2: Ne

      ··

··   Ne   ··   8e

      ··

Ionic Bonding

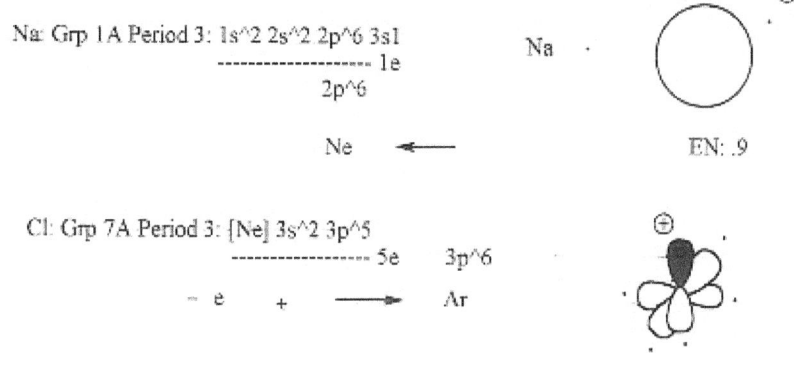

Na: Grp 1A Period 3: 1s^2 2s^2 2p^6 3s1

------------------- 1e

      2p^6

         Ne ←——

Na ·

⊖

EN: .9

Cl: Grp 7A Period 3: [Ne] 3s^2 3p^5

------------------ 5e   3p^6

— e   +   ——→   Ar

⊕

EN: 3.0

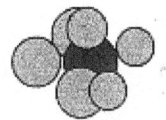 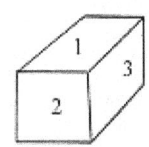

NaCl Crystal: 6 faces of Na or Cl atoms

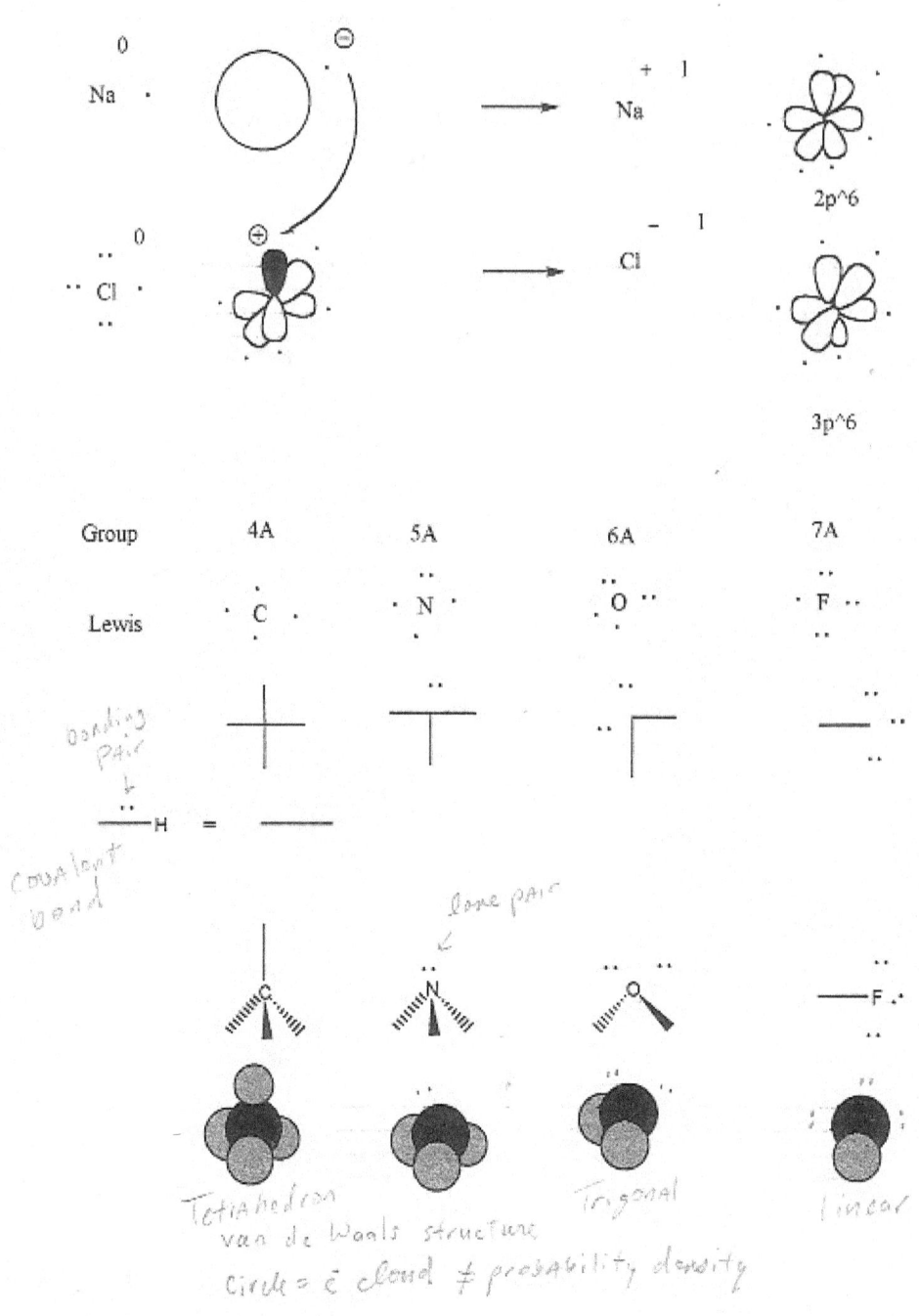

Tetrahedron
van de Waals structure
Circle = ē cloud ≠ probability density

Periodic Table of the Elements
Electronegativity

http://chemistry.about.com
©2010 Todd Helmenstine
About Chemistry

*** Elements > 104 exist only for very short half-lifes and the data is unknown. ***

																	8A
1A																	
1 H 2.20	2A											3A	4A	5A	6A	7A	2 He
3 Li 0.98	4 Be 1.57											5 B 2.04	6 C 2.55	N	O	F	10 Ne
11 Na 0.93	12 Mg 1.31	3B	4B	5B	6B	7B	8B			1B	2B	13 Al 1.61	14 Si 1.90	15 P 2.19	16 S 2.58	Cl	18 Ar
19 K 0.82	20 Ca 1.00	21 Sc 1.36	22 Ti 1.54	23 V 1.63	24 Cr 1.66	25 Mn 1.55	26 Fe 1.83	27 Co 1.88	28 Ni 1.91	29 Cu 1.90	30 Zn 1.65	31 Ga 1.81	32 Ge 2.01	33 As 2.18	34 Se 2.55	Br 2.96	Kr
37 Rb 0.82	38 Sr 0.95	39 Y 1.22	40 Zr 1.33	41 Nb 1.6	42 Mo 2.16	43 Tc 1.9	44 Ru 2.2	45 Rh 2.28	46 Pd 2.20	47 Ag 1.93	48 Cd 1.69	49 In 1.78	50 Sn 1.96	51 Sb 2.05	52 Te 2.1	53 I 2.66	54 Xe 2.6
55 Cs 0.79	56 Ba 0.89	57-71	72 Hf 1.3	73 Ta 1.5	74 W 2.36	75 Re 1.9	76 Os 2.2	77 Ir 2.20	78 Pt 2.28	79 Au 2.54	80 Hg 2.00	81 Tl 1.62	82 Pb 2.33	83 Bi 2.02	84 Po 2.0	85 At 2.2	86 Rn
87 Fr 0.7	88 Ra 0.89	89-103															

Lanthanides

57 La 1.10	58 Ce 1.12	59 Pr 1.13	60 Nd 1.14	61 Pm 1.13	62 Sm 1.17	63 Eu 1.2	64 Gd 1.2	65 Tb 1.2	66 Dy 1.22	67 Ho 1.23	68 Er 1.24	69 Tm 1.25	70 Yb 1.1	71 Lu 1.27

Actinides

89 Ac 1.1	90 Th 1.3	91 Pa 1.5	92 U 1.38	93 Np 1.36	94 Pu 1.28	95 Am 1.3	96 Cm 1.3	97 Bk 1.3	98 Cf 1.3	99 Es 1.3	100 Fm 1.3	101 Md 1.3	102 No 1.3	103 Lr

Legend: 1.0 1.3 1.6 1.9 2.2 2.5 2.8

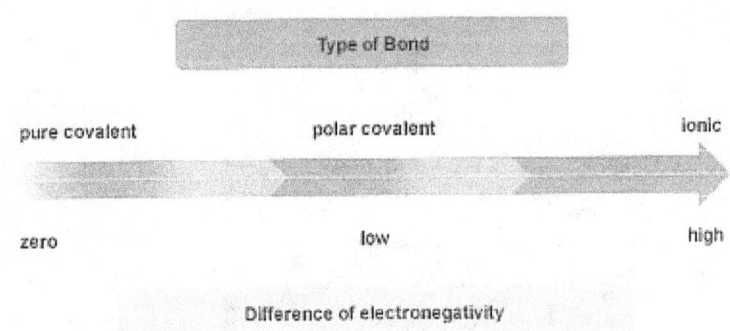

Bond length: distance between the nuclei of 2 atoms joined by a covalent bond.
A function of atom type and bond order (single, double, or triple).

Bond Energy: Amount of energy to break or form 1 mole of covalent bonds between 2 atoms in a molecule in the gas phase.

Delta H = Delta H bonds broken + Delta H bonds formed.
H = enthalpy = heat = energy.

Averaging is involved.

Bond Lengths and Bond Energies		
	Bond Length	Bond Energy
	(nm)	(kJ/mol)
H–H	0.074	435
H–Cl	0.127	431
Cl–Cl	0.198	243
H–C	0.109	414
C–Cl	0.177	328
C–C	0.154	331
C=C	0.134	590
C≡C	0.120	812
C–O	0.143	326
C=O	0.120	803
C≡O	0.113	1075
N–N	0.145	159
N=N	0.125	473
N≡N	0.110	941

Linear Combination of Atomic Orbitals-Molecular Orbitals :

It describes the shape of the orbital and e density distribution.
The molecular orbital is derived from the atomic orbital's overlapping.
Since the orbitals are traveling waves. You can get constructive and destructive behavior.
Constructive create bonding orbitals with a (-) character. They have a larger amplitude.
Destructive create non-bonding orbitals with a (+) charge character.

Constructive = in-phase overlap
Destructive = out-of-phase overlap

1 A.O. + 1 A.O. → 2 M.O. = bonding + non-bonding.

Electrons choose the lowest energy state = bonding orbital.
The energy diagram shows this pattern.

Sigma bond is the s shell bond. It has rotational symmetry, meaning it circular in a cross-section of the bond.

Sigma = bonding orbital.
Sigma* = non-bonding orbital.

The optimal bond length is reach at the lowest energy state.
Bond length is proportional to the overlap.

Bond dissociation energy is the amount of energy needed to break a covalent bond.
Bond dissociation energy is proportional to chemical reactivity. The less energy means more reactive.
Homolytic breakage means each atom receives 1 e from the break. 1 covalent bond = 2 e.

#1: COS(θ)

Ψ : wave equation

Traveling wave moving to the right, and + .

#2: $|COS(θ)|^2$

#3: $|COS(θ)|^2$

Probabillity density (P) : orbital energy, location of e, = 1 (100%). Represented by the area under the curve.

#4: $|COS(θ)|^2 + |COS(θ)|^2$

100% overlap, Amplitude larger, in phase, constructive = bonding, e occupied.

#5: COS(θ − π)

wave moving to the right

#6: COS(θ + 1·π)

wave moving to the left

Phase = 180 ° = 100% overlap = 100% cancellation. P = 0%
Areas cancel. Area(+) = Area (−) added = 0.

#7: $COS\left(θ + \dfrac{1}{2}·π\right)$

90 ° out of phase

#8: $\left|COS\left(θ + \dfrac{1}{2}·π\right)\right|^2$

parital cancellation : determined by the lowest, stable, energy state.

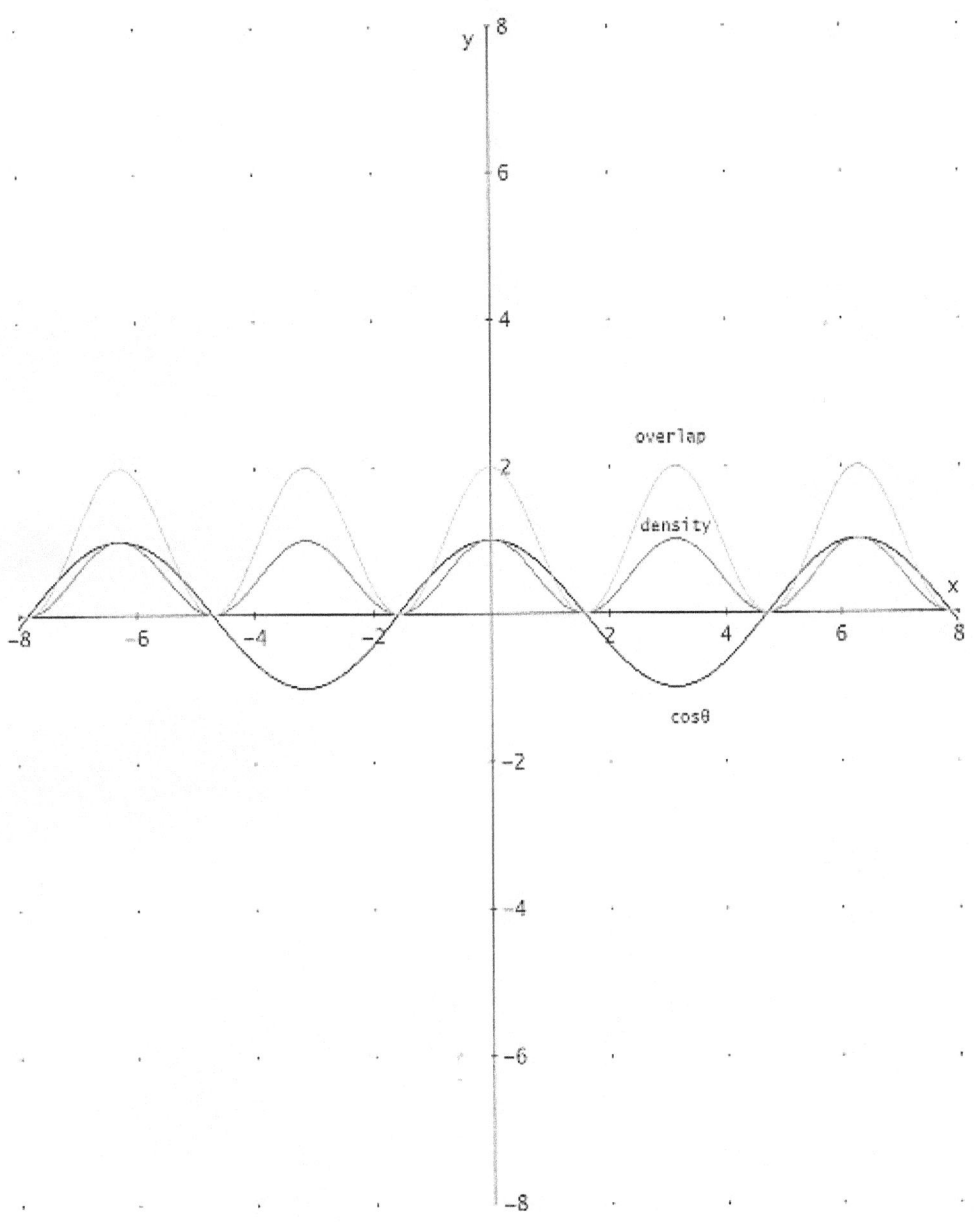

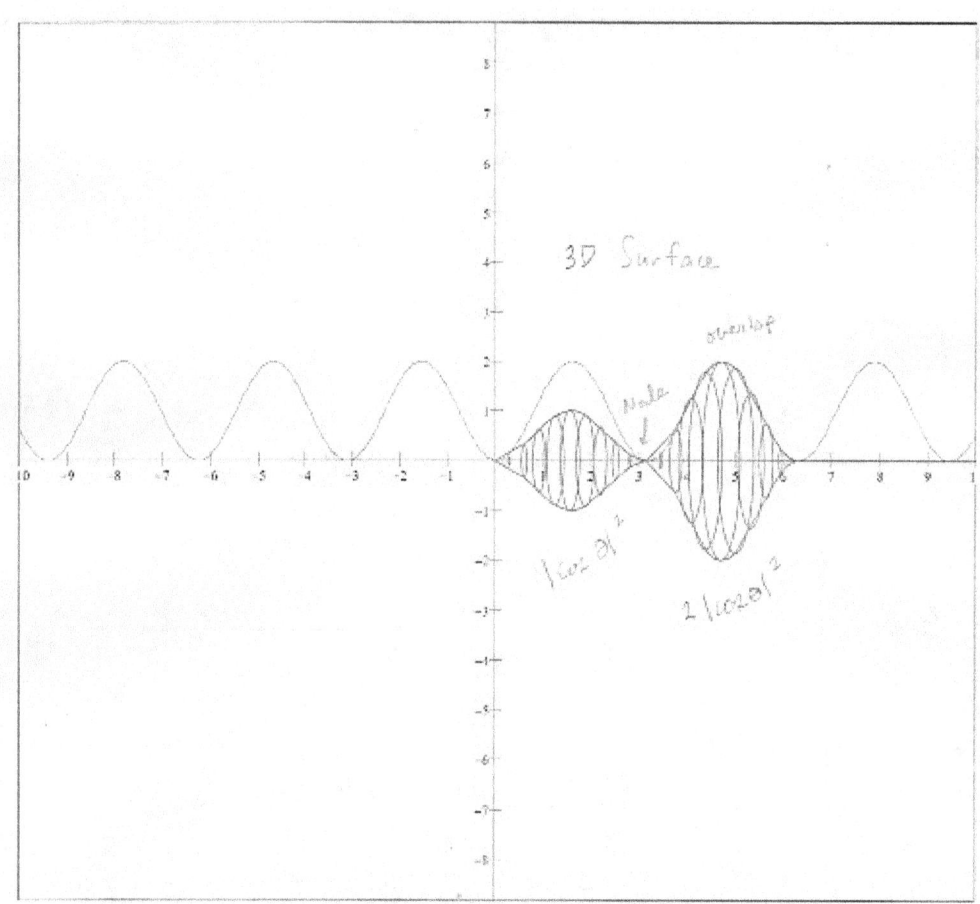

Rotation noted.

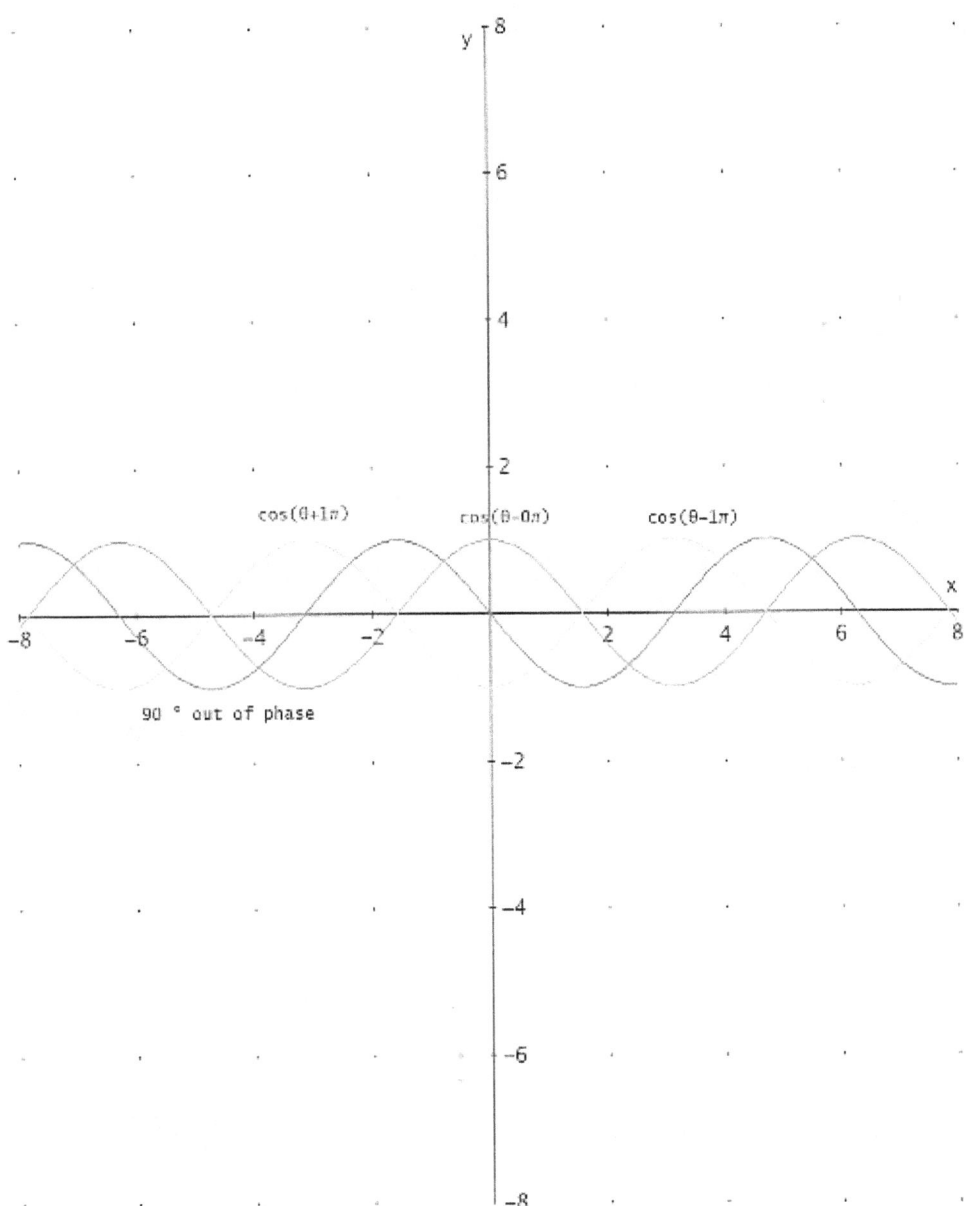

cos(θ+1π)   cos(θ−0π)   cos(θ−1π)

90 ° out of phase

137

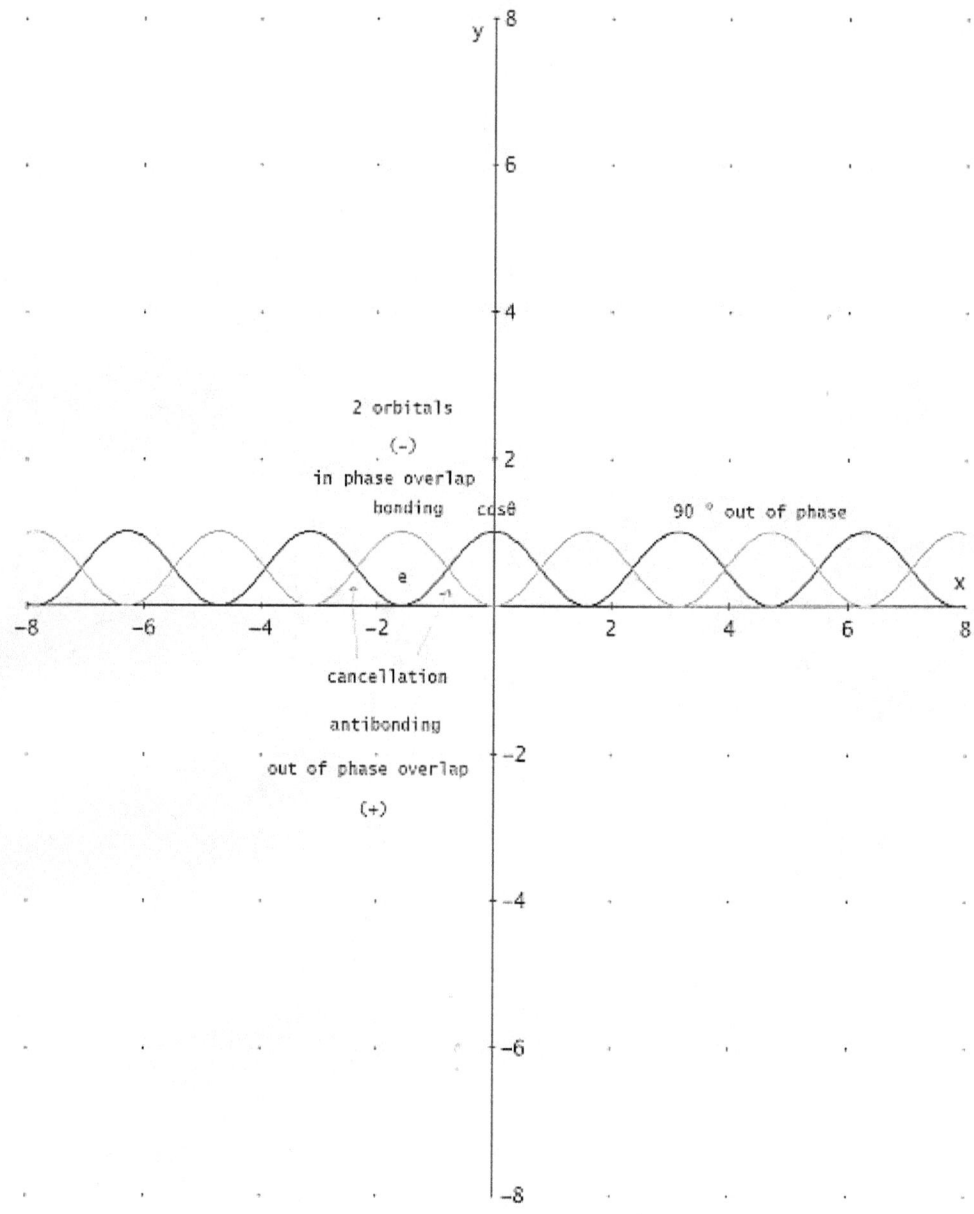

138

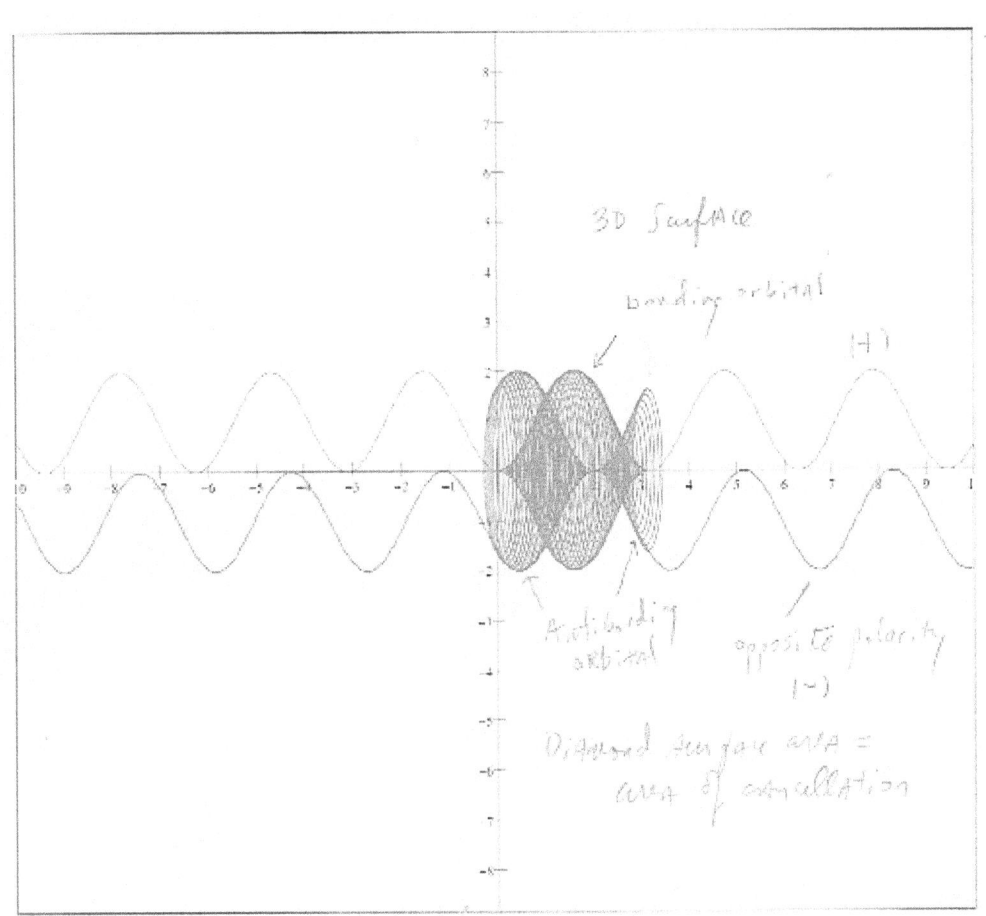

Rotation demonstrated.

The surface shows the cancellation areas in 3D.

Note the non-bonding areas near the nucleus and the outer most distance.

The beauty of the physical nature of the covalent bond.

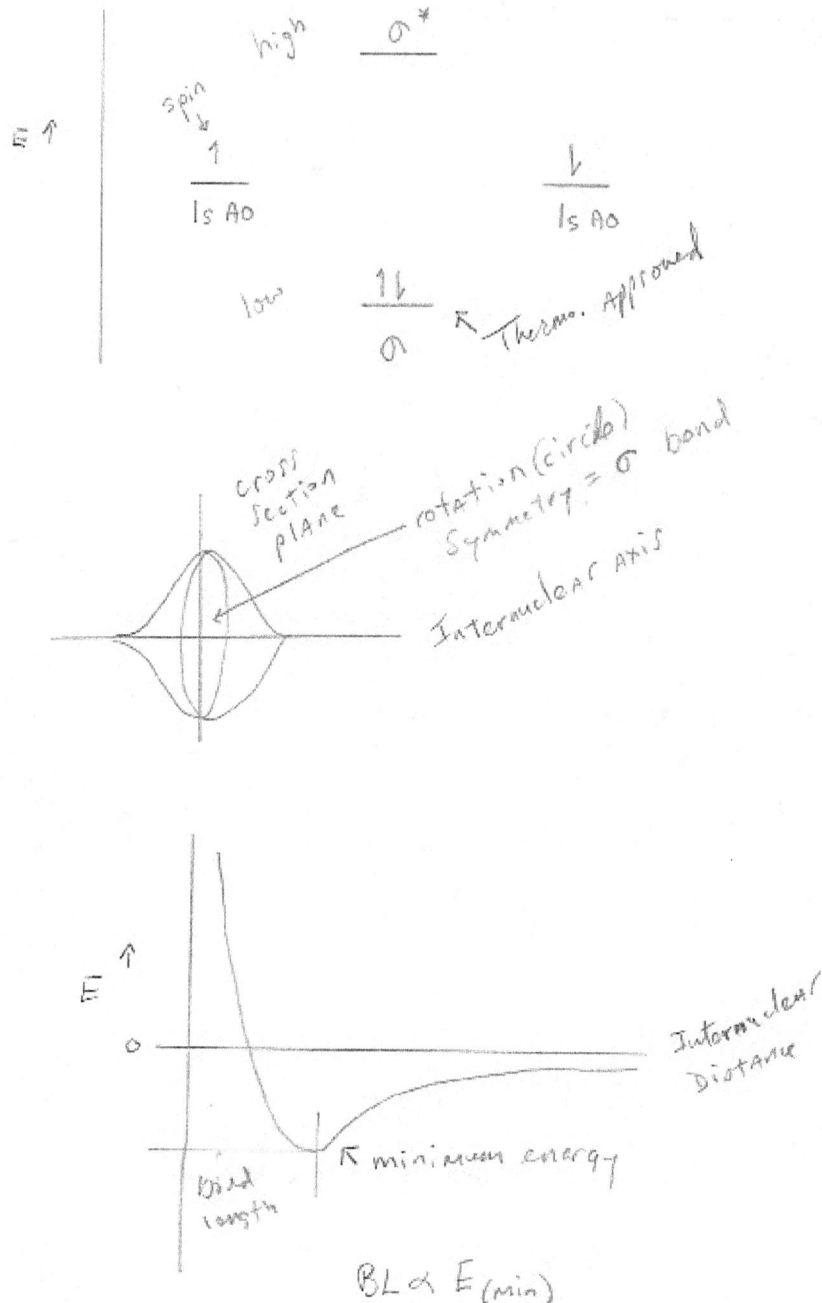

$E \uparrow$

spin high $\quad \underline{} \ \sigma^*$

$\underline{\uparrow} \atop \text{Is AO}$ $\qquad \underline{\downarrow} \atop \text{Is AO}$

low $\quad \underline{\uparrow\downarrow} \atop \sigma$ $\ \nwarrow$ Thermo. Approved

cross
section
plane

rotation(circle) bond
Symmetry = σ

Internuclear Axis

$E \uparrow$

O

Internuclear
Distance

$\nwarrow$ minimum energy

bond
length

$BL \propto E_{(min)}$

Bond	Dissociation Energy, kcal/mole	Bond	Dissociation Energy, kcal/mole
H–H	104	H—F	136
F—F	37	H—Cl	102
Cl—Cl	57	H—Br	87.5
Br—Br	46	H—I	71.3
I—I	36	CH_3—H	103
C—F	108	CH_3CH_2—H	98
C—Cl	81	$(CH_3)_2CH$—H	94.5
C—Br	68	$(CH_3)_3C$—H	91
C—I	55.5	C≡C—H	102
C—O	90	C═C—H	125
C═O	257	C—C	88
O—H	105	C═C	163
		C≡C	200

Note the low energy of the halogens, C-O, O-H bonds, and C-C bonds. Good for chemical reactions.

Orbital Hybridization:

A mathematical operation based on quantum mechanics that explains the geometry of a molecule.

The VSEPR model predicts the geometry of a molecule by arranging all orbitals at maximum distance from each other.

We will now discuss the use of VSEPR to predict the geometry of the molecule and the Valence Bond and Molecular Orbital Theories to further describe the structure of molecules.

Geometry means shape, 3D structure. Atomic nucleic are joined by straight lines. These lines are associated linearly or bent, angular.

The precise geometry is determined by experiment. The shapes can be predicted by the VSEPR Method.

VSEPR=Valence-Shell Electron-Pair Repulsion

Electron-Group: Groups repulsing other groups of electrons.

single e   · 1 Dot

lone pair e   : 2 Dots

single bond   — 1 horizontal line

double bond   = 2 horizontal lines
triple bond   3 horizontal lines

Electron-group geometries:

2 groups = linear
3   trigonal planar
4   tetrahedral
5   trigonal bipyramidal
6   octahedral

VSEPR Notation: AX#E#
A = central atom
X = terminal atom
E = lone pair e
# subscript = number of them

Lewis structure should give the same electron-group geometry.

If no lone pairs present, molecular = electron-group geometry (Identical)
If present, they are not the same and electron-group geometry determines molecular geometry.

Hybridization is the blending of the s and p orbitals. There are 3 types:

sp = 50% s + 50% p:: Triple bond

sp^2 = 33% s + 66% p: Double bond

sp^3 = 25% s + 75% p: Single bond

s orbit has a strong e attracting ability.
p orbit has more e density along the internuclear axis.

Hybridization creates more bonds and they are more stable and stronger. The hybrid orbit has an unsymmetrical e density which allows greater overlap of orbitals.

Sigma bonds are created between hybrids or hybrids and s orbitals.

Hybridization means Mixing: In the case of C, it means the s + p
atomic orbitals

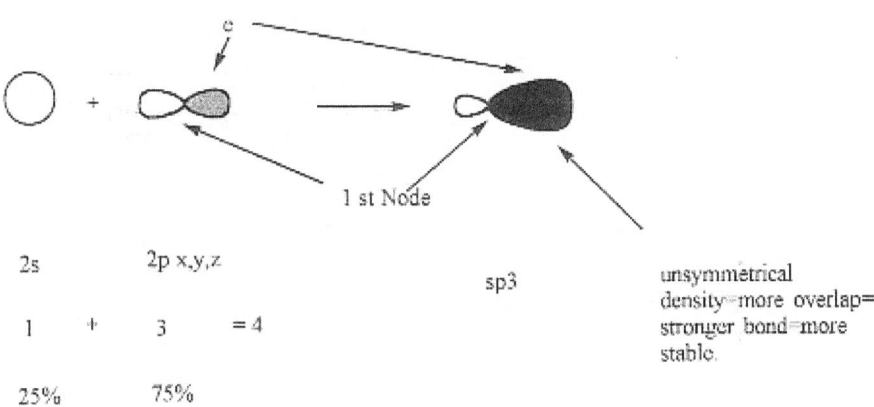

2s   2p x,y,z   sp3   unsymmetrical
                                                             density=more overlap=
1   +   3   = 4   stronger bond=more
                                                             stable.
25%   75%

e is a function of the Wave equation.
The orbit is the electron probability density.
1 st Node is where the nucleus of the atom is.
Node has 0 probability of finding the e.
If the origin of the graph is at the 1st Node, then the Wave equation has a negative part
directed along the -x axis.

Energy Diagram

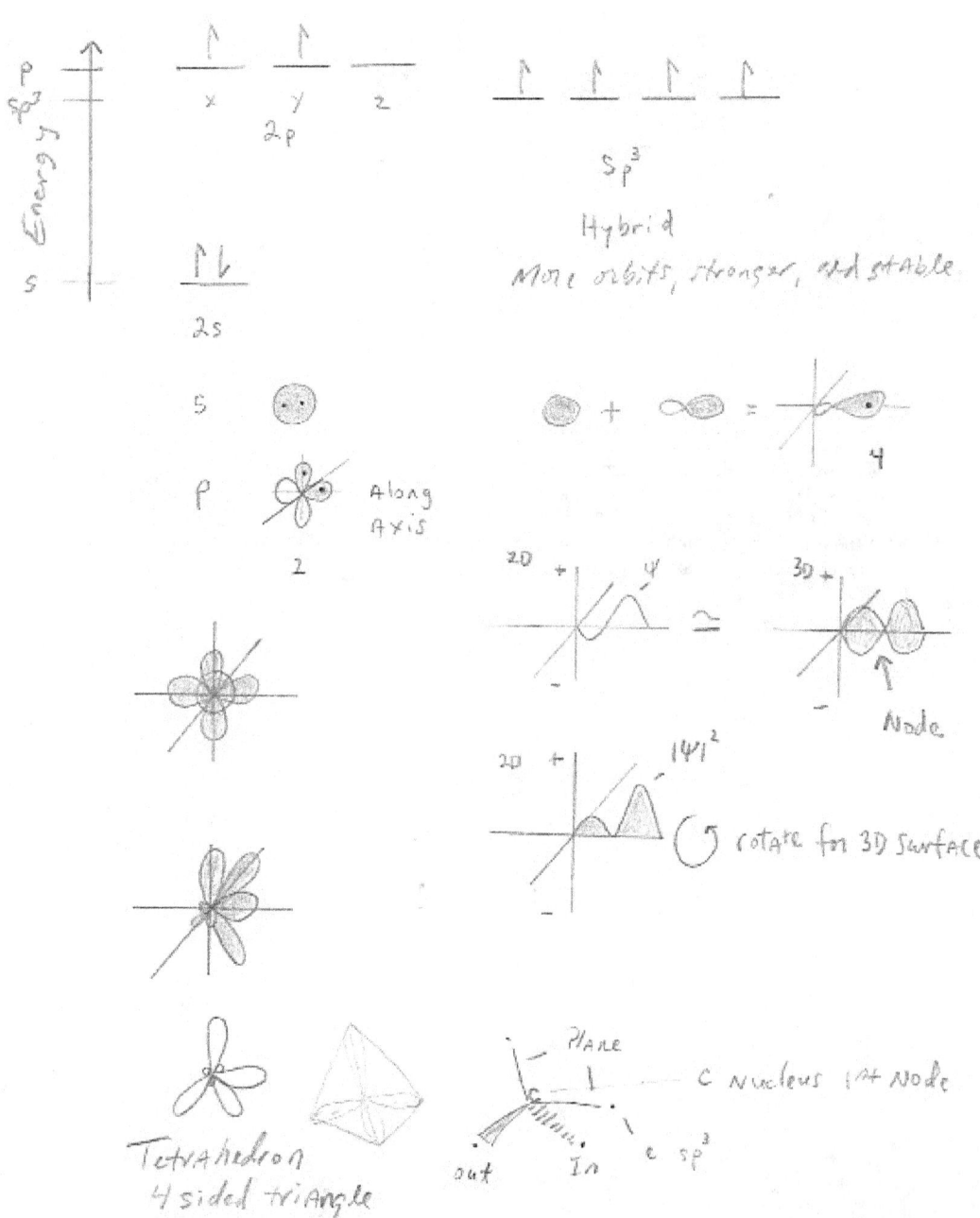

sp^3

Hybrid

More orbits, stronger, and stable

Along Axis

Node

$|\psi|^2$

rotate for 3D Surface

Plane

C nucleus 1st Node

out   In   C sp^3

Tetrahedron
4 sided triangle

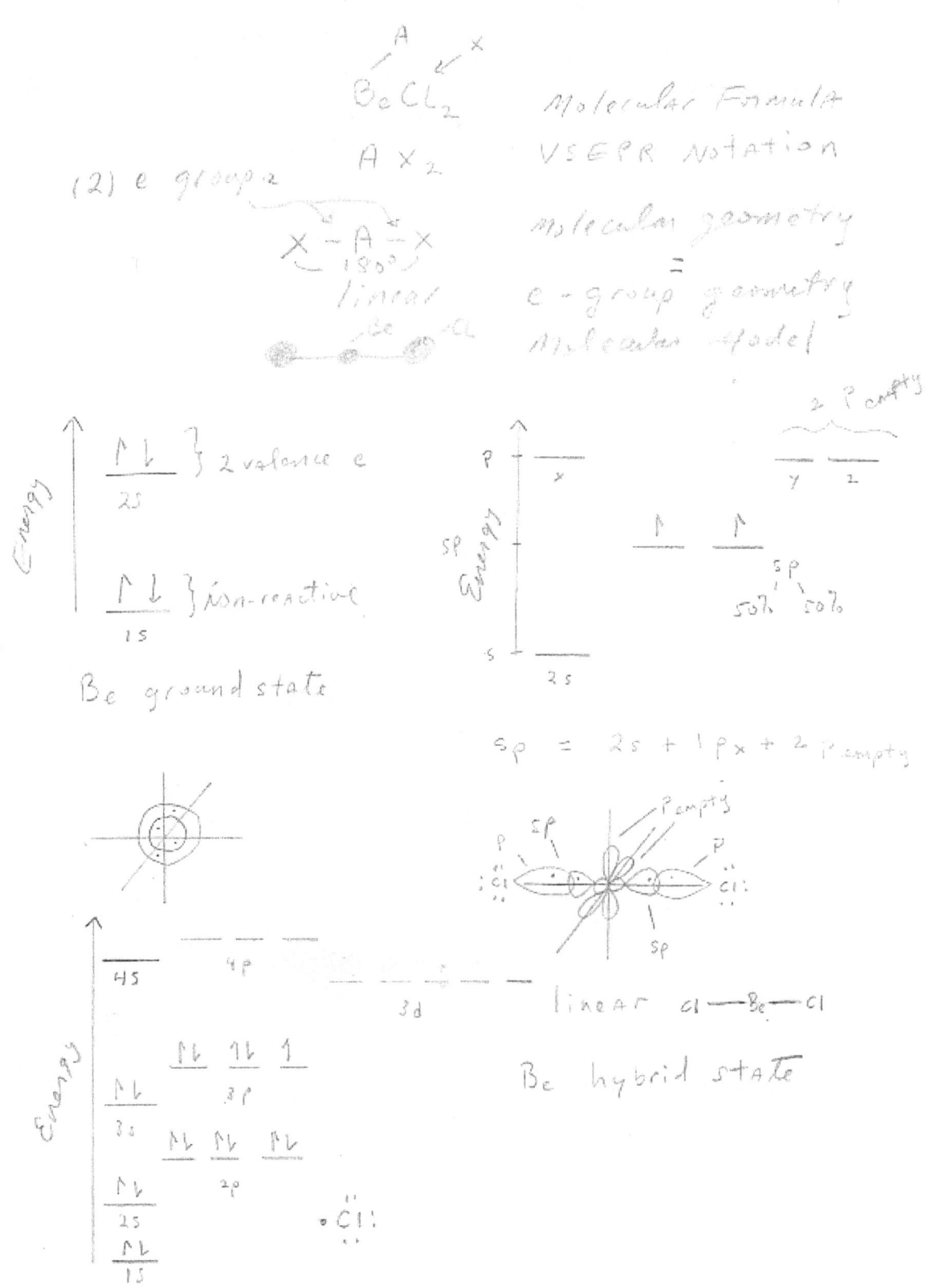

$BeCl_2$ Molecular Formula

$A X_2$ VSEPR Notation

(2) e groups

molecular geometry
=
e - group geometry

$X - A - X$
$180°$

linear

Molecular Model

Be—Cl

2 P empty

↑↓ } 2 valence e
$2s$

↑↓ } non-reactive
$1s$

Be ground state

P   x   y   z

sp   ↑   ↑

sp
50%   50%

s $2s$

$sp = 2s + 1p_x + 2 p empty$

P empty

P   sp   P

:Cl: :Cl:

sp

linear   cl — Be — cl

Be hybrid state

$4s$ $4p$

       $3d$

↑↓   ↑↓   ↑
$3p$

↑↓
$3s$

↑↓   ↑↓   ↑↓
$2p$

↑↓   :Cl:
$2s$

↑↓
$1s$

Note: Be ground state is spherical. It's hybrid state is directed along the x axis. In this state, the electron attraction is more directed, larger and stronger. In a sphere, the electric field radiates from a very large surface, which weakens the strength of the field. By directing the field along a straight line, the field strength increases, making bonding possible.

Also, the Cl ground state has a semi-vacant 3p orbital. 1 e will fill this orbital, stabilizing Cl.

Be converted from a 2 valence e, 2s stable orbit to 2 sp orbits which are higher in energy and need to be stabilized (reactive for bonding).

Note also the 2 p empty orbitals which could potentially accept 2 lone pair e groups.

Notes:

BF3   ~~Molecular Formula~~

AX3   VSEPR

3 e groups

e group geometry   Trigonal Planar

Molecular geometry   Trigonal Planar

120*

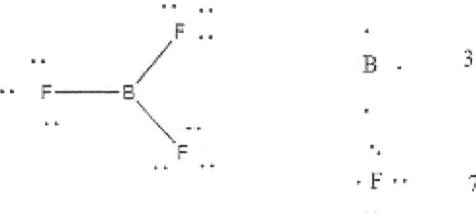

B : Group 3 = 3 valence e, 5 total e

F : Group 7=7 valence e, 9 total e

B = 2s + 2px + 2py + 2pz empty -> 3 sp^2 hybrid

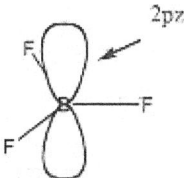

2pz

B + F all in the same plane.

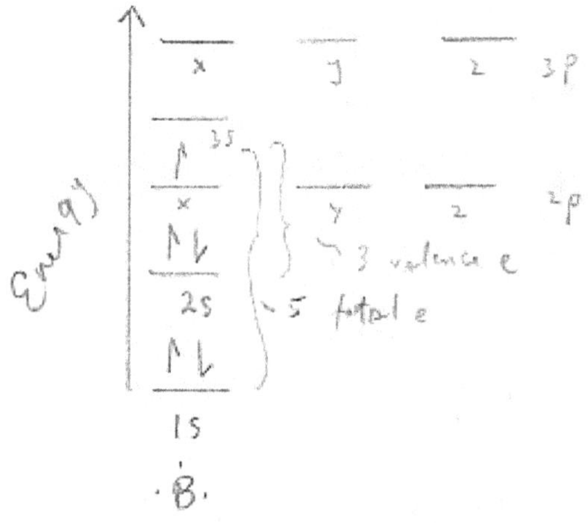

Energy

x   y   z   3p

3s

x   y   z   2p

3 valence e

5 total e

2s

1s

·B·

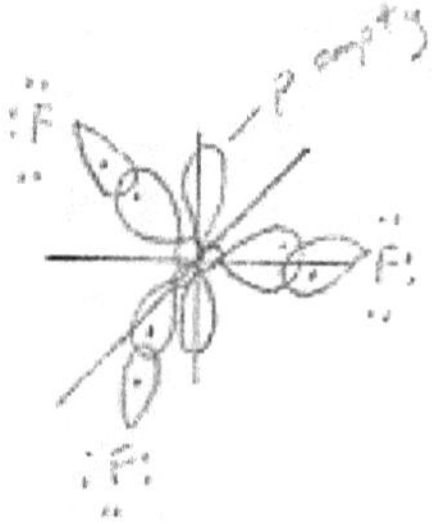

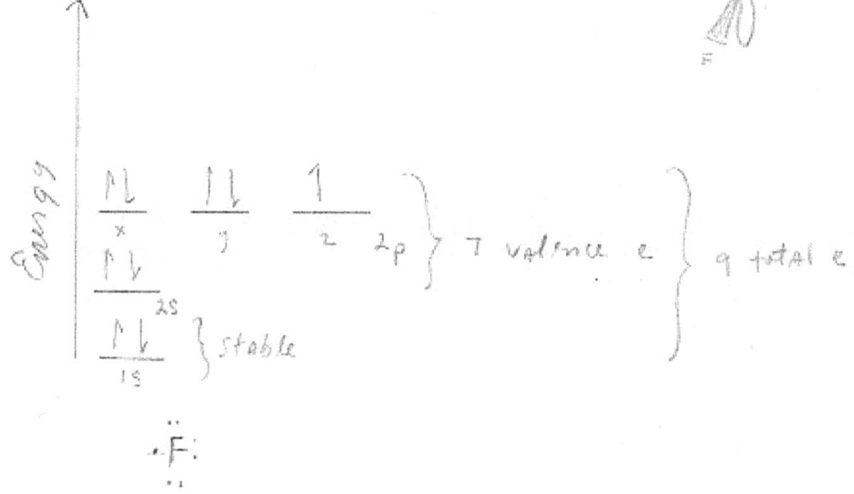

Energy

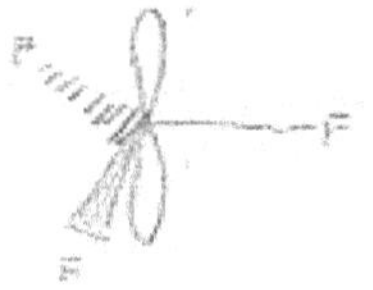

x

y   z   2p   7 valence e   9 total e

2s

stable

1s

·F̈:

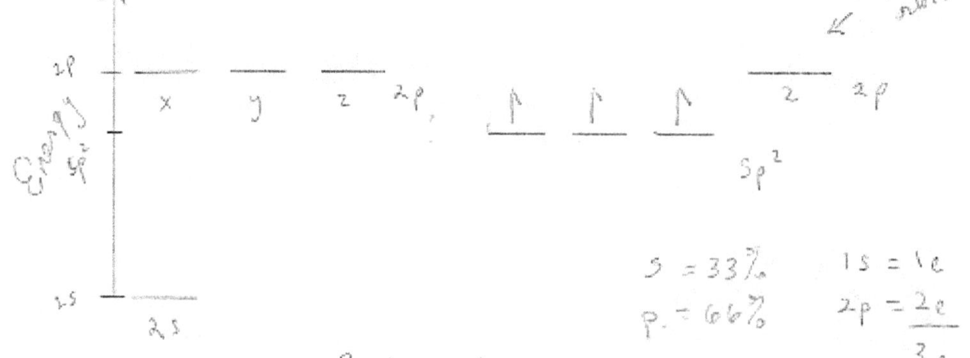

Energy

sp²

2p   x   y   z   2p₁   z   2p

empty? orbit

sp²

2s   2s

B hybrid

$s = 33\%$ $1s = 1e$

$p = 66\%$ $2p = \dfrac{2e}{3e}$

SO2	MF
AX2E	VSEPR
3	# e grps
1	# lone pairs
120*	angle

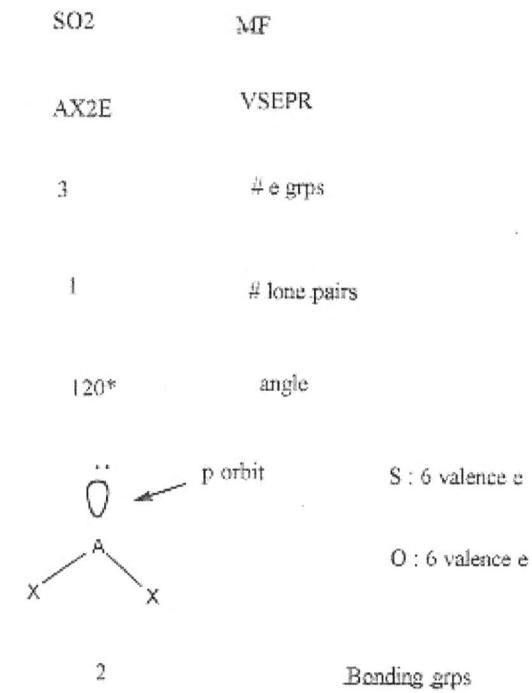

S : 6 valence e

O : 6 valence e

| 2 | Bonding grps |

The Octet rule for S is not satisfied. Need 2 e.

Resonance from either O fixes the deficit.

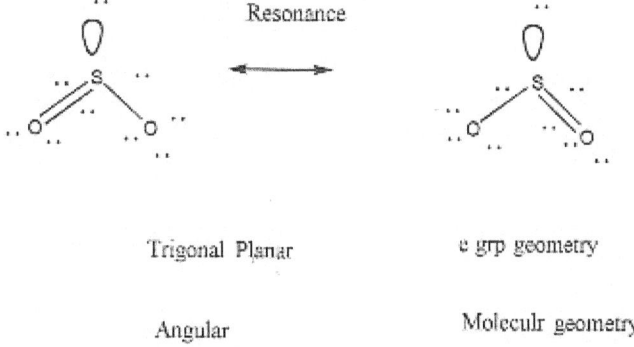

Resonance

Trigonal Planar   e grp geometry

Angular   Moleculr geometry

e grp geometry does not equal molecular geometry

Molecular geometry is atom based, not electron.

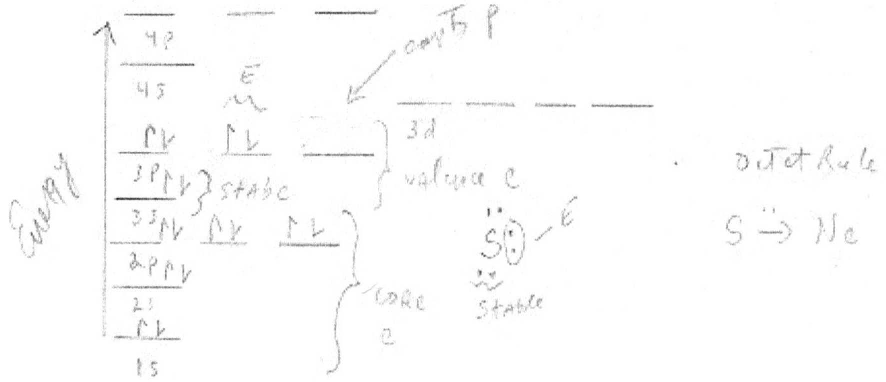

Energy

4p

4s   E

3p ⇅⇅ ⇅⇅ } 3d

      } stable   valence e

3s ⇅⇅ ⇅⇅ ⇅⇅ }

2p ⇅⇅

2s ⇅⇅   } core e   S̈   E   stable

1s

empty p

Octet Rule

S ⇌ Ne

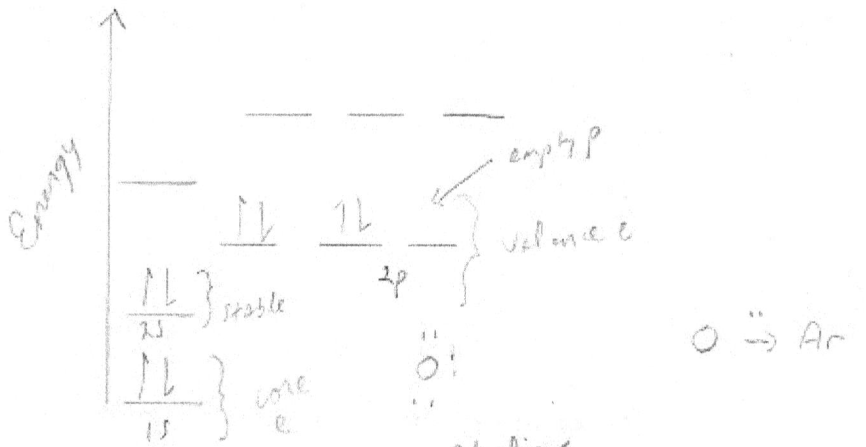

Energy

⇅⇅ ⇅L   empty p

              valence e

⇅⇅ } stable   2p

2s

⇅⇅ } core   Ö

1s   e

O ⇌ Ar

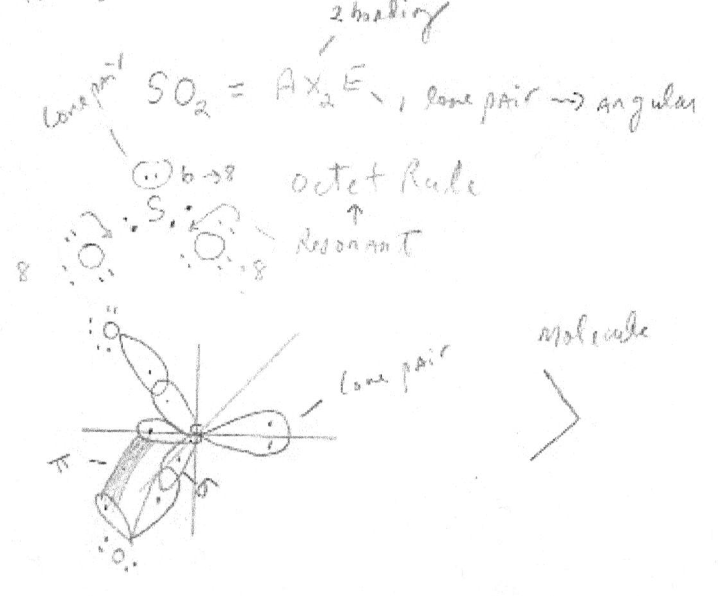

                    2 bonding

lone pair SO₂ = AX₂E , lone pair → angular

        6→8   Octet Rule

   S   ↑
8 Ö   Ö 8   Resonant

                lone pair   Molecule

π   σ

S and O look electronically alike. They both need 2 electrons to stabilize.

There are 3 atoms: 1 S and 2 O. The shape looks Linear. The lone pair places a restriction on the model.

Hybridization of sp^2 type can manage the bonding model.

The Lewis structural analysis helps along with the Octet Rule.

Resonance saves the model.

The model is angular and planar, the boomerang.

This is a good example of the use of the periodic table, Lewis structure, Octet Rule, and Resonance.

The reactivity of the molecule is evident:

1. The lone pair on S can act as a nucleophile: attacking a p+.
2. On the O side of the molecule is a resonating (-) field of charge. This too can act as a nucleophile and attack a p+.

We could expect acid-base reactions, oxidation-reduction reactions, and nucleophilic attack.

CH4	MF
AX4	VSEPR
4	# of e groups
0	# of lone pairs
109.5*	Angle
sp^3	Hybridization : 4 Bonds
4	Bonding groups
4	C: # valence e
1	H: # valence e
Sigma bond	C: sp^3 - H: s orbits.
Tetrahedral	e group geometry
Tetrahedral	Molecular geometry

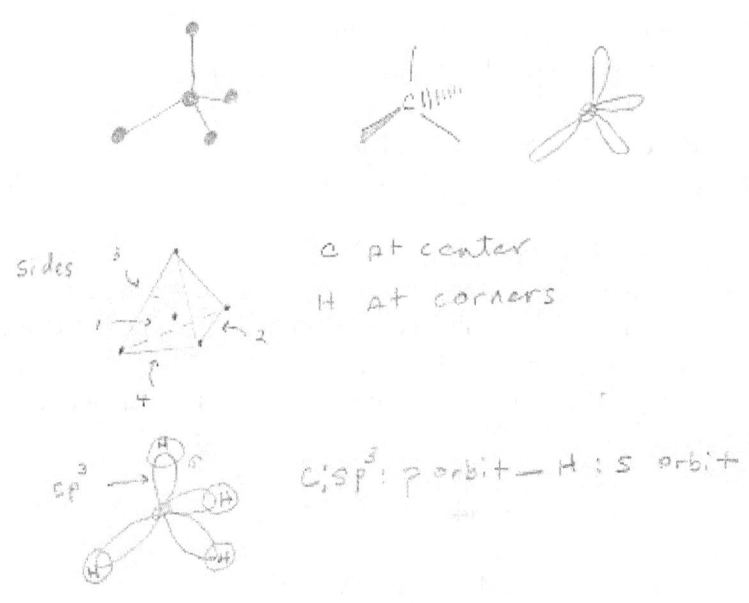

sides

e at center
H at corners

sp³ →

c:sp³: p orbit — H: s orbit

NH3	MF
109.5*	Angle (Degrees)
AX3E	VSEPR
4	# e grp
Tetrahedral	e grp geometry
Trigonal Pyramidal	Molecular geometry
1	# lone pair

Pyramidal = vertical orientation

Trigonal Pyramidal = Tetrahedral without vertical apex

vertical pyramid

Notes:

OH2	MF
4	# e grps
2	# lone pairs
AX2E2	VSEPR
Tetrahedral	e grp geometry
Angular	Molecular geometry
109.5*	Bond Angle

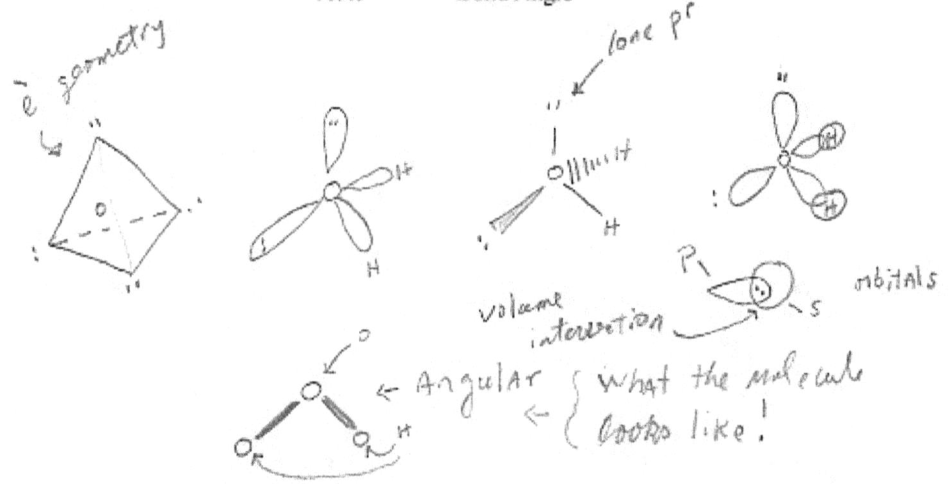

e⁻ geometry

lone p⁻

O

H

volume interaction

P

S

orbitals

← Angular { what the molecule looks like!

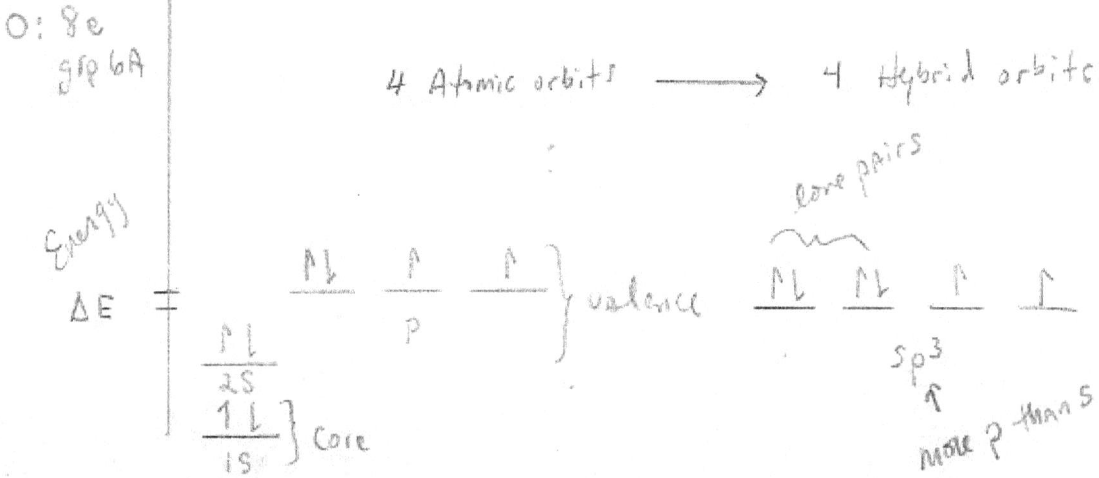

O: 8e
grp 6A

Energy

ΔE

4 Atomic orbits ⟶ 4 Hybrid orbits

lone pairs

↑↓   ↑   ↑ } valence   ↑↓ ↑↓ ↑   ↑
        P   sp³
↑↓
2S   ↑
↑↓ } core   more P than S
1S

H : 1e
grp 1A

Energy

p

1S

Notes:

PCl5	MF
5	# e grps
0	# lone pairs
AX5	VSEPR
Trigonal bipyramidal	e grp geometry
Trigonal bipyramidal	Molecular geometry
90*, 120*, 180*	Bond Angle

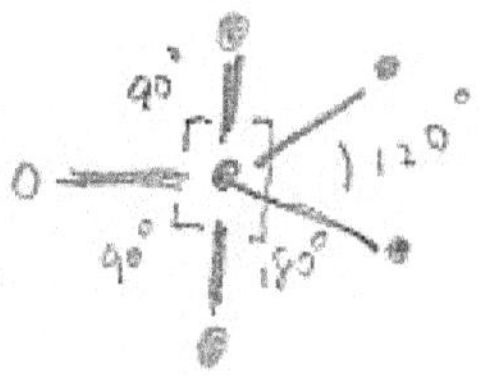

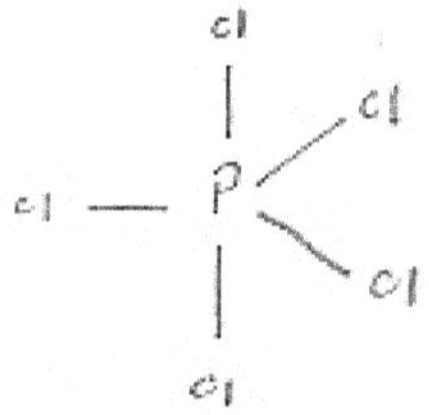

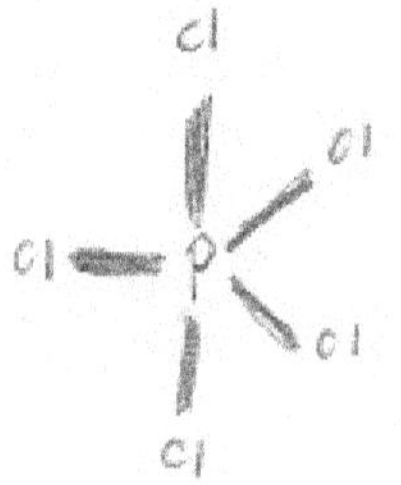

trigonal
bipyramidal

sp^3d

Mol. geometry

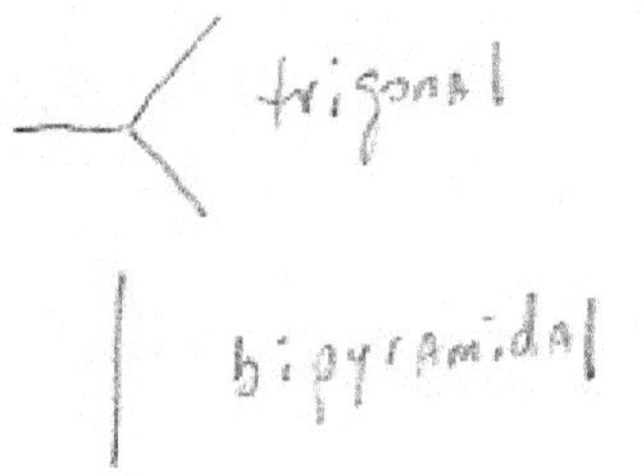

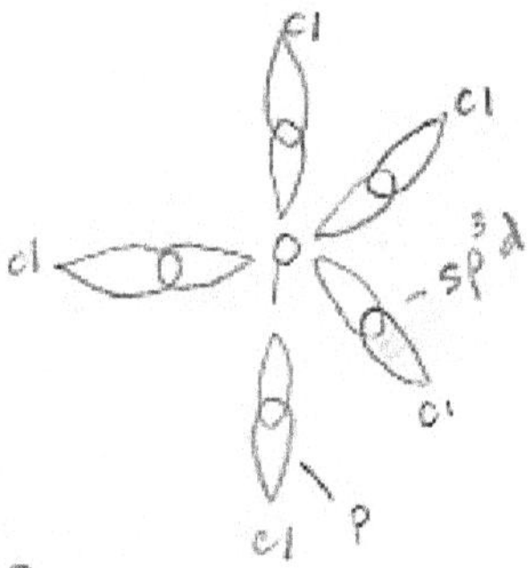

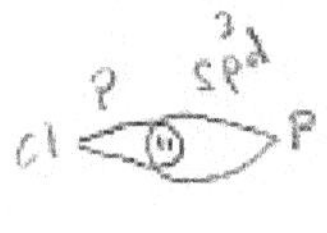

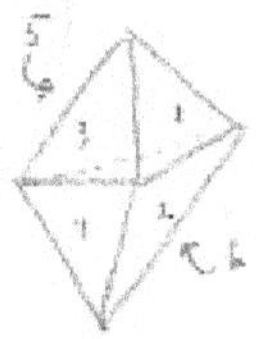

e geometry

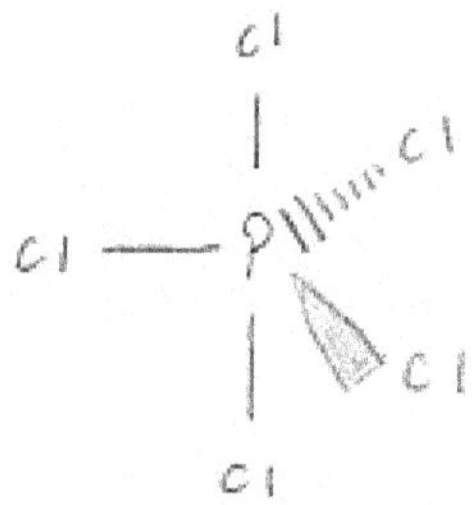

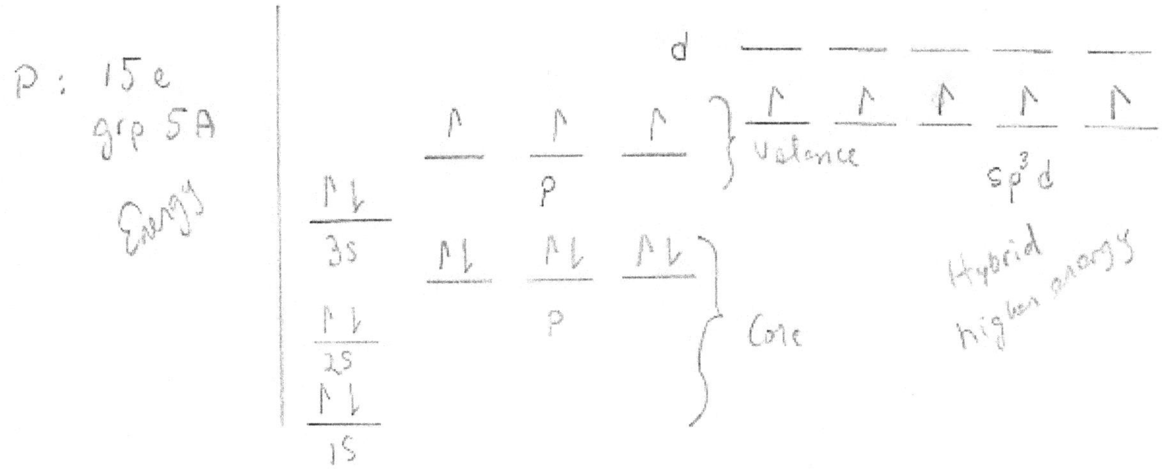

P : 15 e
grp 5A
Energy

d

3s ↑ ↑ ↑ } Valence
       P

↑↓
3s

↑↓ ↑↓ ↑↓ } Core
       P

↑↓
2s
↑↓
1s

↑ ↑ ↑ ↑ ↑ sp³d

Hybrid
higher energy

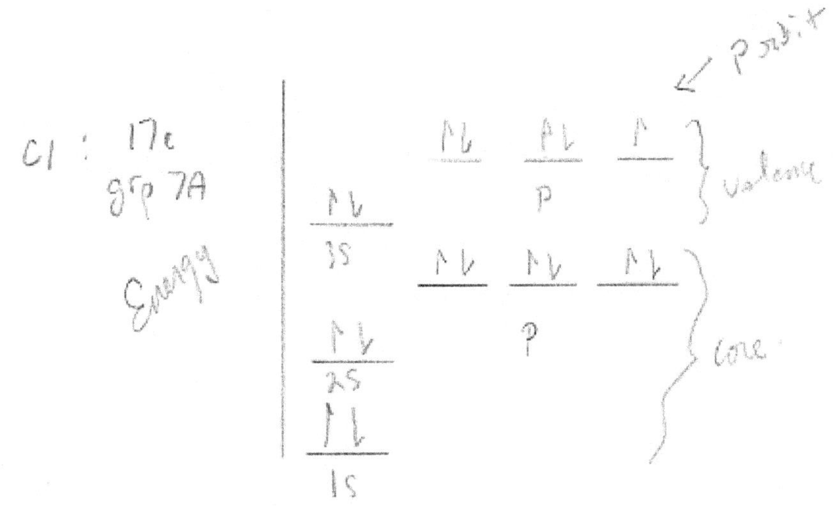

← P orbit

Cl : 17 e
grp 7A
Energy

↑↓ ↑↓ ↑ } valence
       P

↑↓
3s

↑↓ ↑↓ ↑↓ } core
       P

↑↓
2s
↑↓
1s

Notes:

SF4	MF
5	# e grps
1	# lone pairs
AX4E1	VSEPR
Trigonal Bipyramidal	e grp geometry
Seesaw	Molecular geometry
90*,120*, 180*	Bond Angle

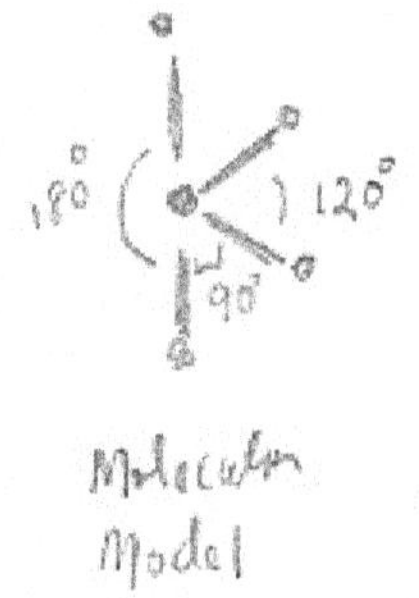

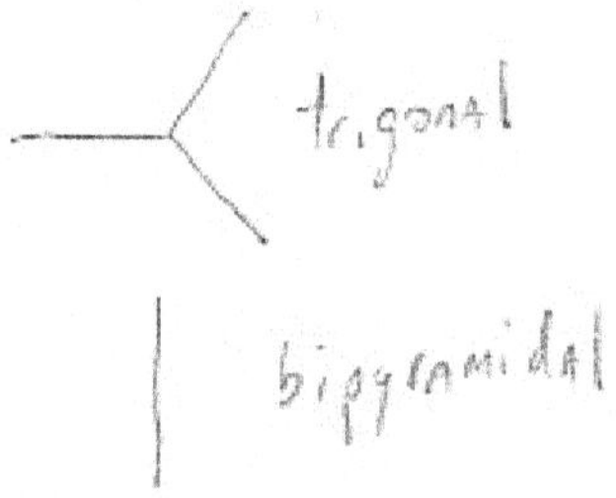

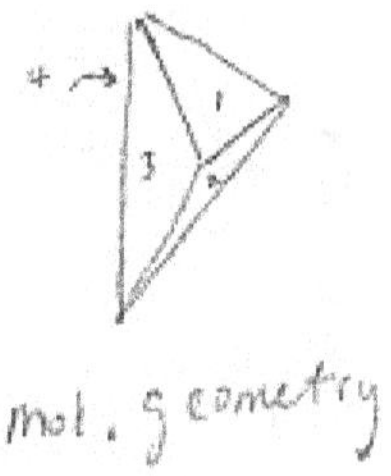

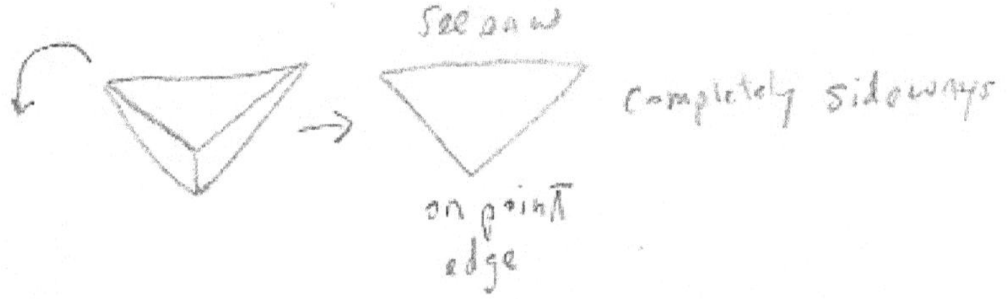

Molecular Model

180° ( ) 120°
90°

trigonal

bipyramidal

Fl
|
S — Fl
|
Fl
|
Fl

lone pr
: — S — Fl
Fl
|
Fl
|
Fl

è geometry

mol. geometry

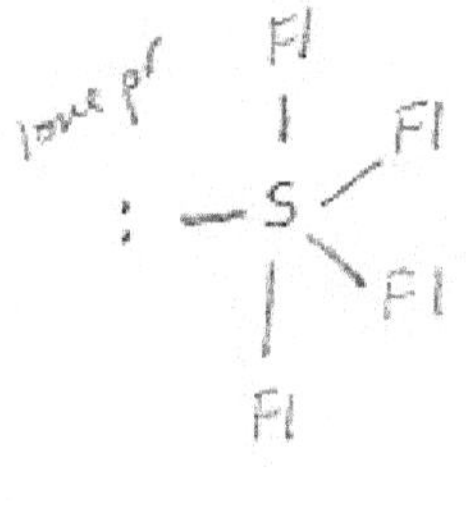

Fl
|
: — S ||||''' Fl
| ◁ Fl
Fl

Seesaw

on point
on point edge

completely sideways

S: 6A 16e

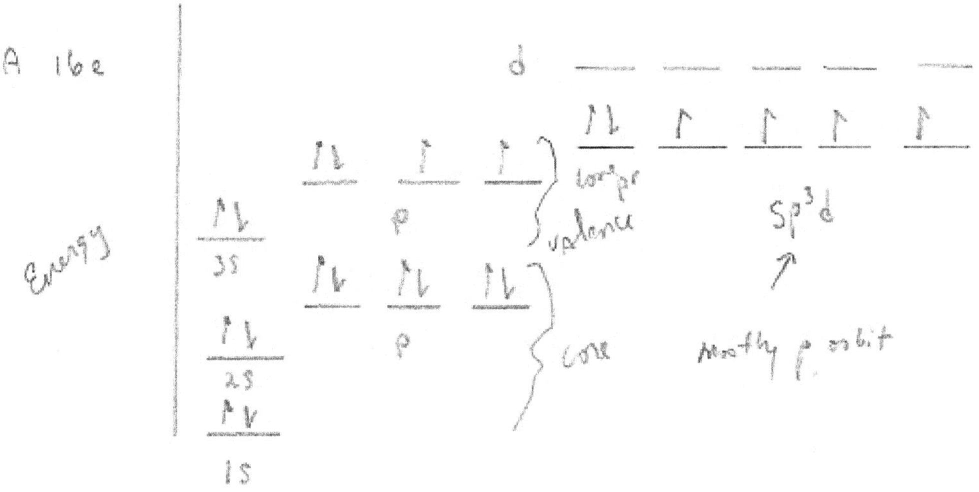

Energy

d

↑↓ ↑ ↑ ↑ ↑
sp³d
↑ mostly p orbit

↑↓ ↑ ↑ } lone pr valence
3s   p

↑↓ ↑↓ ↑↓ } core
p

↑↓
2s

↑↓
1s

Fl: 7A 9e

Energy

↑↓ ↑↓ ↑
p

↑↓
2s

↑↓
1s

p   sp³d

sp³d   Fl   p
Fl
: S
Fl
Fl

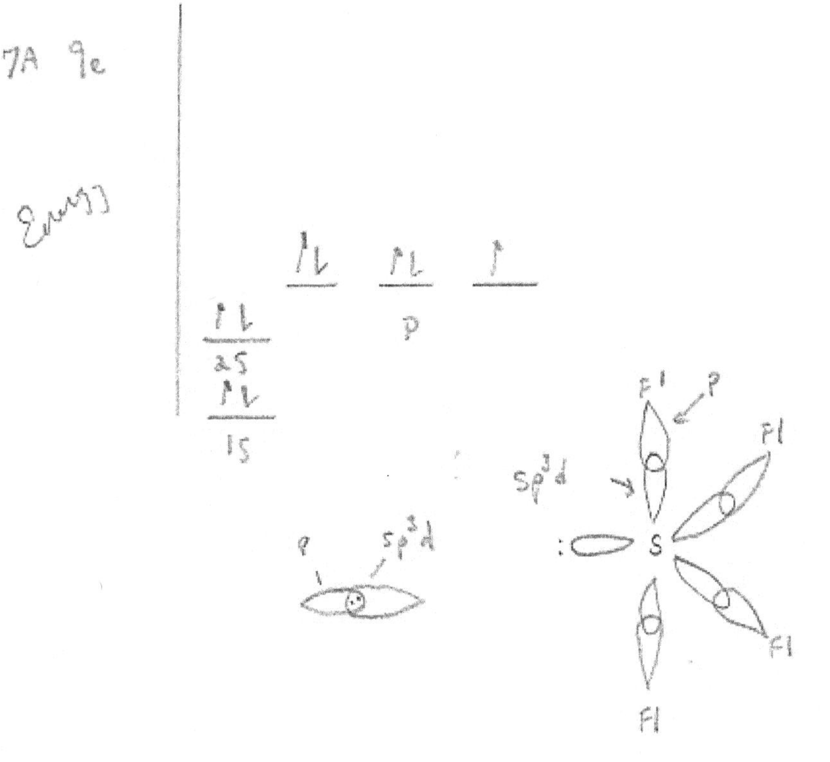

ClF3	MF
5	# e grps
2	# lone pairs
AX3E2	VSEPR
Trigonal Bipyramidal	e grp geometry
T-shaped	Molecular geometry
90*, 180*	Bond Angle

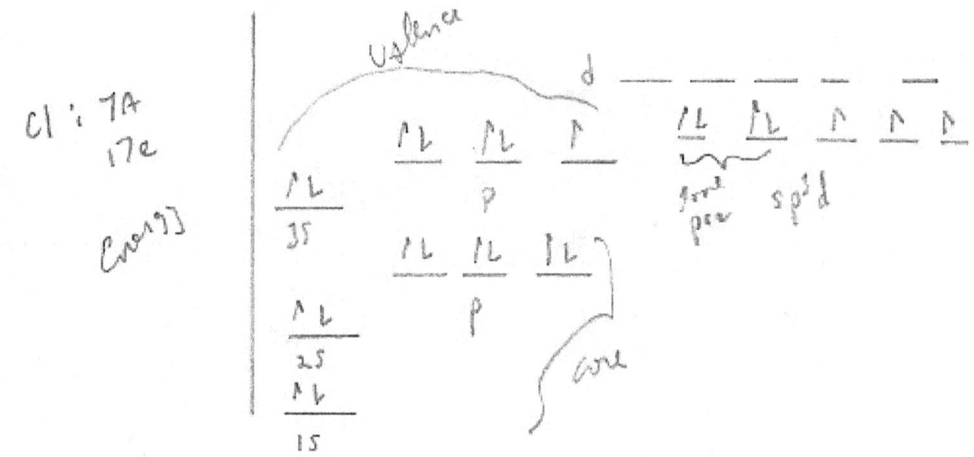

Cl : 7A
17e
(Ne 13)

valence
d

see above for Fl : p orbit

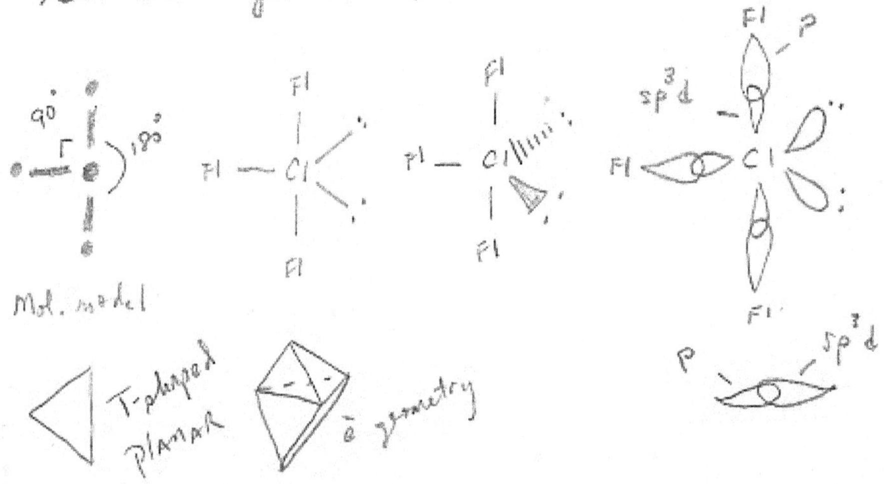

Mol. model

T-shaped
Planar

e geometry

sp³d

XeF2	MF
5	# e grps
3	# lone pairs
AX2E3	VSEPR
Trigonal Bipyramidal	e grp geometry
Linear	Molecular geometry
180*	Bond Angle

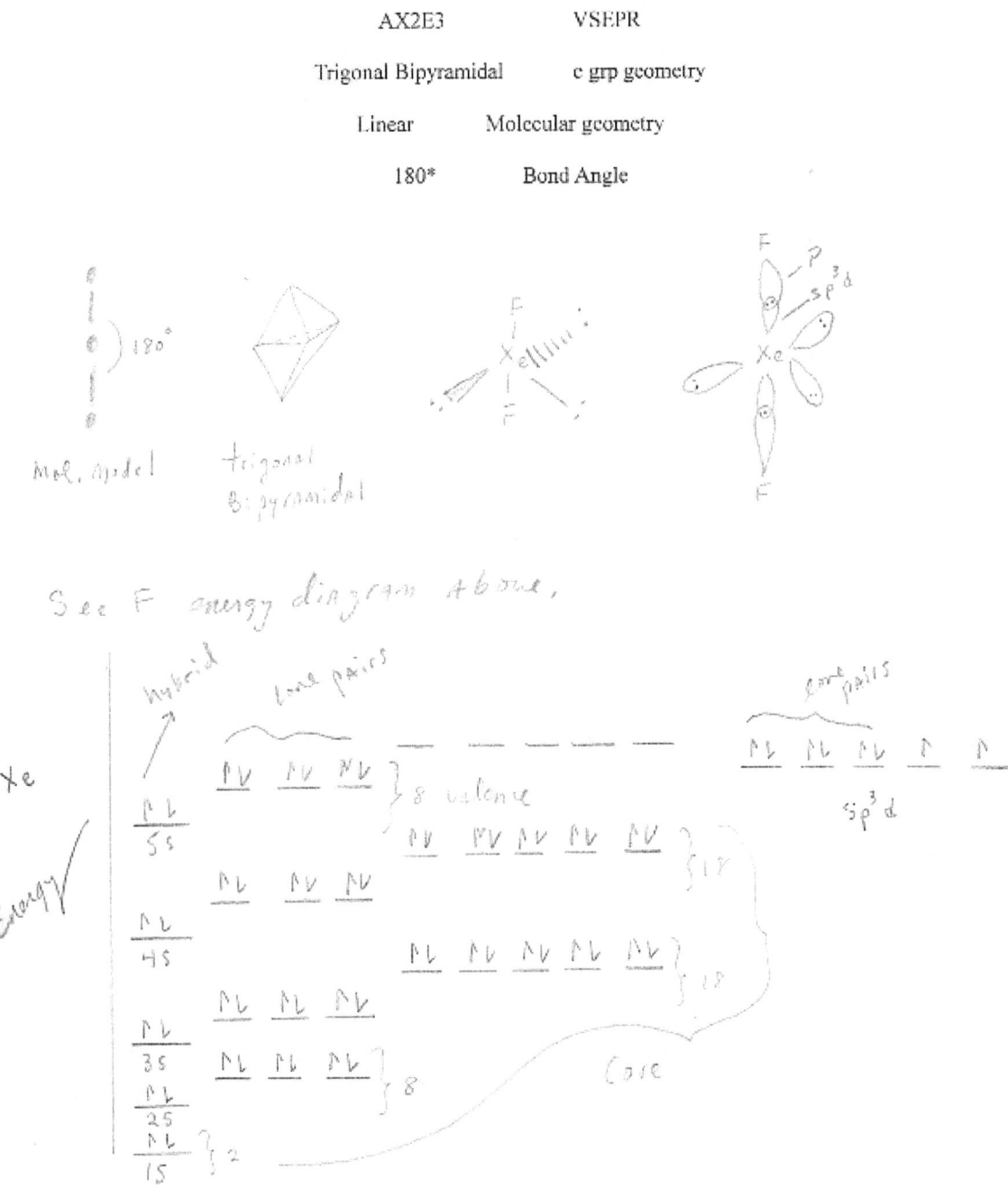

See F energy diagram above.

160

SF6	MF
6	# e grps
0	# lone pairs
AX6	VSEPR
Octahedral	e grp geometry
Octahedral	Molecular geometry
90*, 180*	Bond Angle

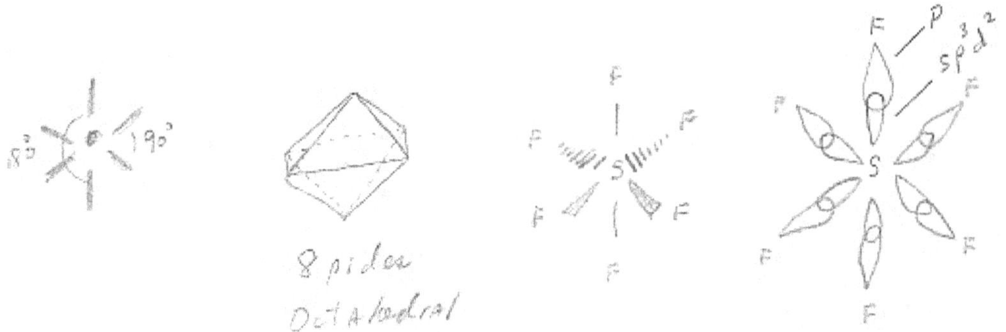

8 pides
Octahedral

See F energy diagram above.

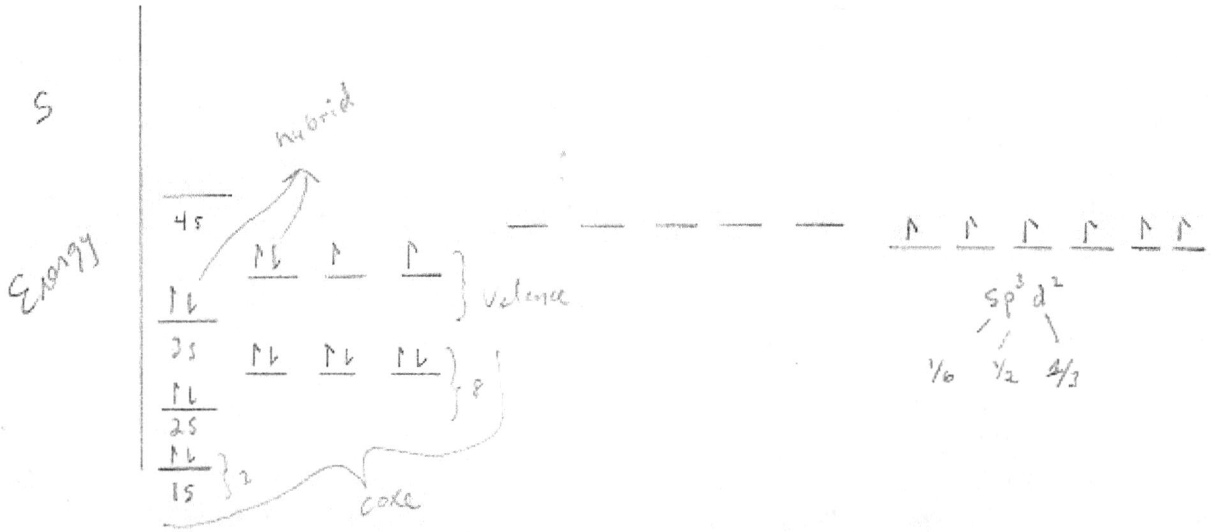

BrF5	MF
6	# e grps
1	# lone pairs
AX5E1	VSEPR
Octahedral	e grp geometry
Square Pyramidal	Molecular geometry
90°	Bond Angle

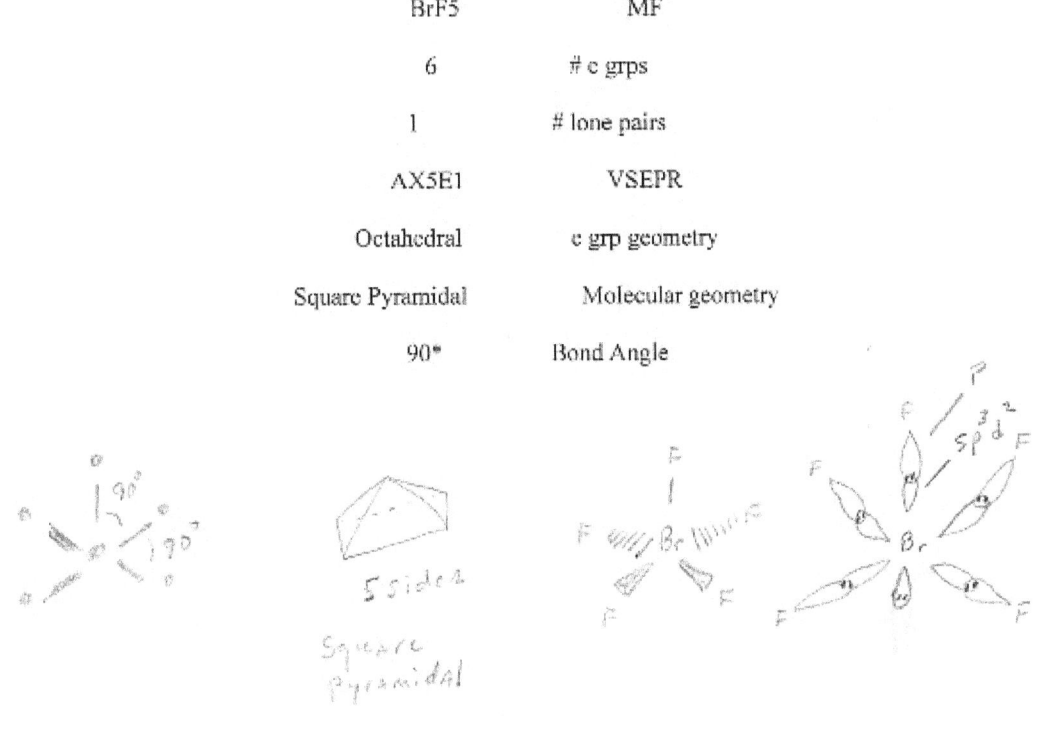

5 sides

Square Pyramidal

See F energy diagram Above

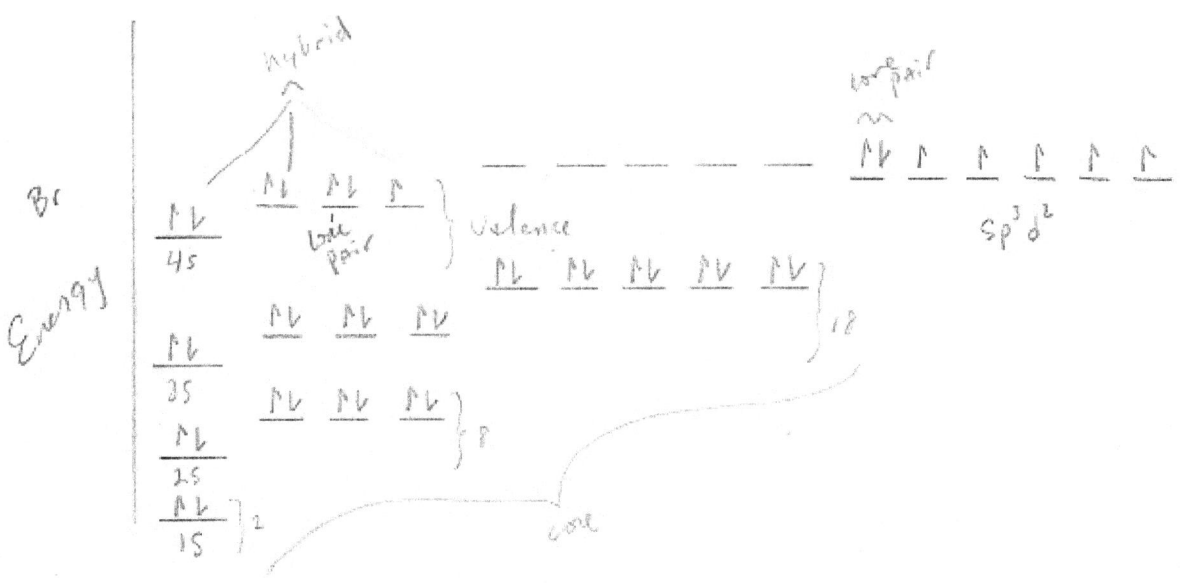

XeF4	MF
6	# e grps
2	# lone pairs
AX4E2	VSEPR
Octahedral	e grp geometry
Square Planar	Molecular geometry
90*	Bond Angle

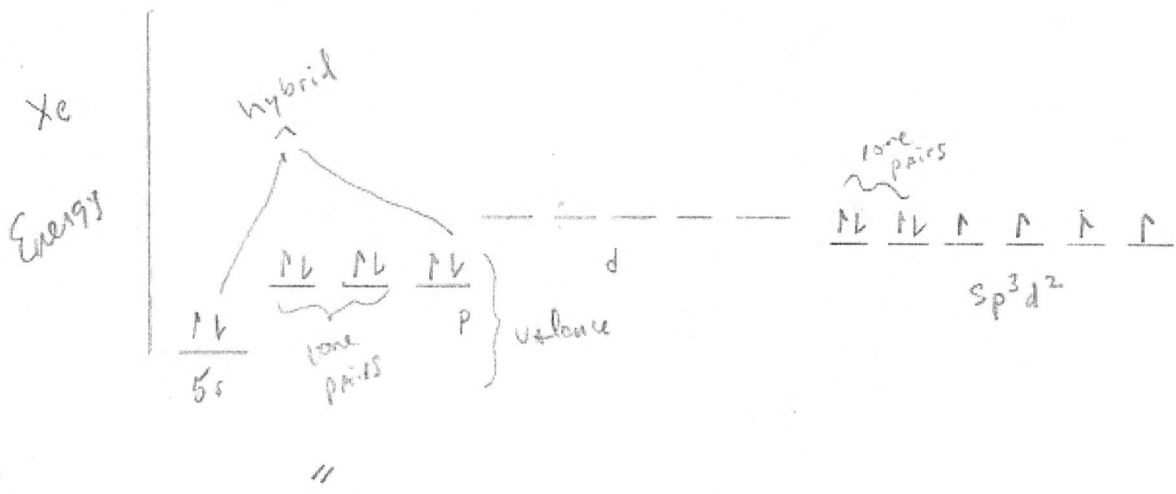

2 sides

See F energy diagram above.

VSEPR is a method used to try to predict the shape of a molecule.
Use the following steps:

1. Draw Lewis structure. This accounts for all the valence electrons for each atom. In each group of atoms there is only 1 central atom which hybridizes. Remember each atom adheres to the Octet Rule for stabilization. I used the energy diagram of the

central atom to visualize the hybridization process. This way you can see which atomic orbits contribute to the hybrid. Formal charges and resonance does not alter the prediction of the model.. Both only alter the electron number and position of lone pairs and multiple bonds.

2. Count the bonds and lone pairs around the central atom.
3. What is the electron group geometry? This is electron group based.
4. What is the molecular geometry? This is atom based.

In the above examples, there is much more information given to tie the the relative concepts together. In the VSEPR method, the Lewis structure, Octet Rule, Bonding and Lone pairs are utilized to predict the geometry of the molecule. The bond angles are predicted in this manner. Using linear lines to generate planes is an attempt to create a 3D approximated image of the molecule. Molecules are more spherical in reality according to Quantum Mechanics.

The ultimate model is chosen based on the lowest potential energy configuration.

e repulsion is greatest to lowest in the following order:

90* > 120*>180*

In the example that follows, the nitrate ion, pay close attention to to valence electron # of each atom, covalent electron # of the bonding atoms, and the total electron #.

There are 4 atoms: 1 N = central atom and 3 O = bonding atoms.

N = 5 e in valence and O = 6 electrons in valence according to the periodic table.

Draw the 3 covalent bonds between the atoms.

Remember each covalent bond equals 2 electrons: 1 from N and 1 from O.

Fill in the balance of the valence electrons to the central atom and fill in the valence electrons around the bonding atoms: N = 5 with 1 lone pair and each O = 6 + 1 = 7 e.

The total electron count is 24, but the apparent total electron count is 26. You need to subtract the difference of 2 e. The N loses the lone pair.

The 3 O atoms seem to each have a -1 charge. This is resonance. The overall molecule has only 1 (-1) charge. We have 2 floating double bonds occurring 1/3 of the time between 2 O atoms.

Remember VSEPR does not distinguish between Single, Double, or Triple bonds.

If the N maintained it's lone pair, the molecular model would have been trigonal pyramidal with an electron model of tetrahedral. Instead it has a trigonal planar for both models.

VSEPR Method

Determine the Molecular Shape

e pair repulsion / valence shell only

eg. nitrate ion

Molecular Formula : NO3 -

1 N atom : valence = 5 -> 5
3 O atoms : valence = 6->18
-1 = 1e
Total e = 24
from periodic table

N=central atom

O=bonding atoms

Covalent bond=2 e

Octet rule = 8e

Total 26 e
so subtract 2 e

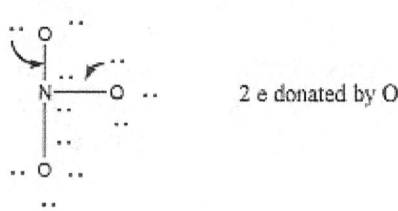

2 e donated by O

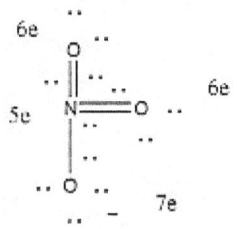

N has 3 e, needs 2 e for 5 valence

O has 1 e more than valence : therefor -1

Single, double, and triple bonds = 1 bonding bond

The -1 floats through the O atoms by resonance

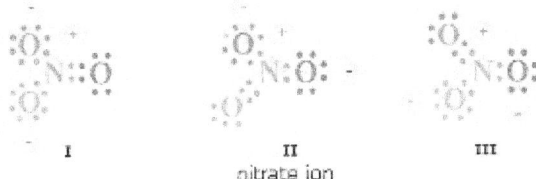

I   II   III

nitrate ion

These are the resonant Lewis structures of the Nitrate ion. You see 2- charges and 1+ charge. We could eliminate 1- and 1+ charge by adding another double bond in the structure. This would not alter the molecular shape using VSEPR method.

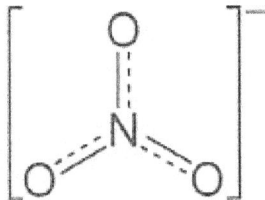

The above figure is the resonant structure of the ion. You see only 1- charge.
This charge is from the whole molecule. Each O atom contributes to this charge.
The dotted line represents a floating colvalent bond. At any 1/3 bond moment, there are 2 double bonds rotating around the nitrogen. The Nitrogen has a 0 charge and 1 of the 3 O has a -1 charge.
N=5 e valence, 6 required for resonance, Net +1 : 6 - 5 = +1
O=6e valence, O in double bond = 6, O in single bond = 7 : 6 - 7 = -1 * 2 = -2

Remember the VSEPR method treats Single, Double, and Triple Bonds as 1 bonding bond.
In a more simple way, since there are no lone pairs, given the N atom as the central atom and the O atoms as the peripheral atoms, there is only one molecular shape : Trigonal Planar.
This is a good example of a resonant structure who's structure is an average at best.

One final comment on this resonant structure. If at 1 point in time, you could see 3 Double Bonds in the structure, the N atom would take on a +1 formal charge. The molecule has a dipole moment enabling the resonance process.

The N atom changes from $0 \rightarrow +1$ and the O atom $0 \rightarrow -1$ formal charge.

VSEPR Method
nitric acid : HNO3

24 valence e = 1+5+3*6

· ·
· N ·
·
· ·
· O · ·
·

H   p +

An acid is a protonated specie. The H+ would chose the O- of Nitrate ion to protonate.

NO3-   ⌒   + H   ⟶   HNO3

Base   Acid

2 central atoms = N + O

N has 3 e grps with AX3 notation
O has 4 e grps with AX2E2

N e grp geometry is trigonal planar
O e grp geometry is tetrahedral

N molecular geometry is trigonal planar : 120*
O molecular geometry is angular : 109*

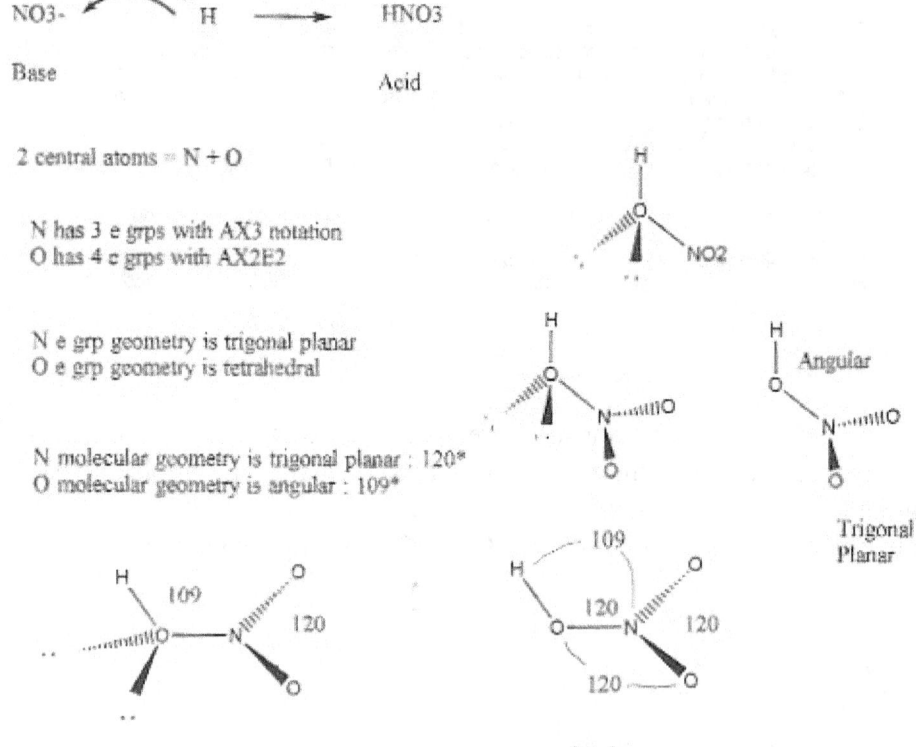

molecule

The total e number does not change because only a p+ was added to the nitrate ion.
The nitrate ion with it's lone pair on the O atom attacked the p+ with a nucleophilic attack
This is an acid base reaction.
The double bond still resonants around the N atom. The net charge is 0.

Despite the different resonant structures, the molecular model predicted by VSEPR holds.

In 1/3 of a time interval, there will be 1 double bond. But if you can sample a larger time interval, say 2/3 or 3/3, could you observe 2 or 3 double bonds? If you could do this, this would not change the prediction using the VSEPR method. During the presence of 3 double bonds, the N atom would have a formal charge of -1. The total charge still would be 0.

The VSEPR only concerns itself with e pairs; whether they are in a covalent bond or alone.

The Lewis structure makes it possible to see chemical reaction capabillity.

Errata: 1. In 2 molecular drawings, the bond that moves away from you is pointing in the wrong direction. It needs to be flipped. 2. The curly arrow between the H+ and nitrate ion needs to be reversed because it is a nucleophilic attack not an electrophilic attack.

VSEPR Method
XeF2

..
·· Xe ··   8   1 atom, central
..

covalent bond

..
· F ··   7   2 atoms, peripheral
..

Total = 2*7+8=22 e

·· ·· ·· ··
·· F——Xe——F ··   No Formal charge: q = 0
·· ·· ·· ··

Octet rule satisfied

Note the 3 pr e grps around Xe

AX2E3

e grp geometry is trigonal bipyramidal

Mol geometry is linear

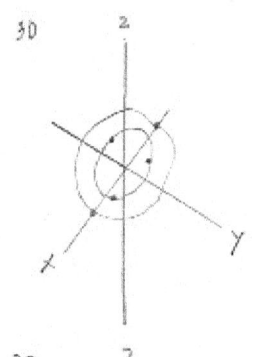

180* | Xe
Linear |
         F

180* Xe

Molecular model using atoms only maximum repulsion. Minimum potential energy.

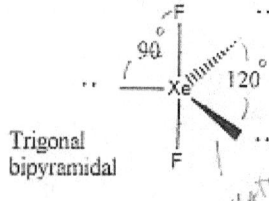

Trigonal bipyramidal

90   120

e grp geometry using e prs only for maximum repulsion. Minimum potential energy,

It is very had to see the relative positions of the atoms in the 3D graphs. But in the 2D graph, it is obvious. These are not drawn to scale. You get the idea. For instance, the two atoms at the top of the x circle need to be farther apart on the circle. The atoms on the x circle are all 120* apart. Don't think that all the atoms are in the same plane. These same two atoms have different positions, 1 away from you and other towards you. In the 3D graphs, at least 1 of the atoms on the y axis, should be on the y axis. Note that the axes can be interchanged because they are arbitrary.

This molecule was easier to model. There is no resonance. The Formal charge is 0.

Dipoles

The covalent bond between to atoms can be polar or non-polr.

Polar means that the 2 atoms have different Electronegativity. EN is the strength that the atom has in wanting the e. eg. H being to the left in the periodic table wants to give up the e. Cl being to the right in the periodic table wants to keep the e. This behavior promotes stabilization of the atom. The atom seeks a minimum potential energy state. Satisfying the Octet rule through bonding, the chemical bond, does just that. Atoms with different EN create a polar bond. In the Lewis structure, it is depicted as +/- delta, which means partial charge. The molecule is therefore polar. The dipole monment (u) is the magnitude of the charge separation. u = delta*d. d=distance in meter. Delta = charge in Coulomb. A polar molecule has a non-zero u. A non-polar molecule has a u = 0. A unit of a u is called the debye (D). 1D = 3.34*10^-30 C*m. The apparatus that measures u acts as a capacitor. It stores a charge. If a polar molecule is measured, it raises the charge capacity. The math symbol used to depict a u is the vector. A straight line that has both direction and magnitude.

Notes:

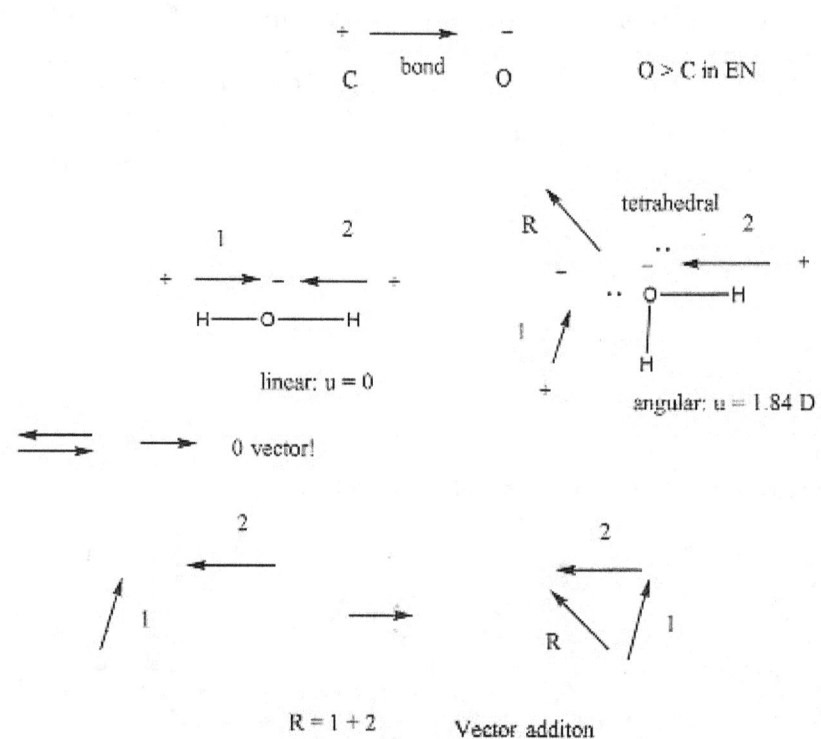

C   bond   O   O > C in EN

linear: u = 0

angular: u = 1.84 D

0 vector!

R = 1 + 2   Vector additon

The utility of dipoles: predicting polarity and dipole moment for compounds. VSEPR is used to predict the shape of the molecule.

Notes:

Which molecule is polar? CCl4   CHCl3

By simple observation: Atoms have different ENs. Only 1 molecule has a H atom. By vector cancellation, it appears the one with the H atom may have a non-zero u. This one is polar.

VSEPR : both AX4 : e : Tetrahedral mol. : Tetrahedral

AXn : n are all the same substituents = non-polar

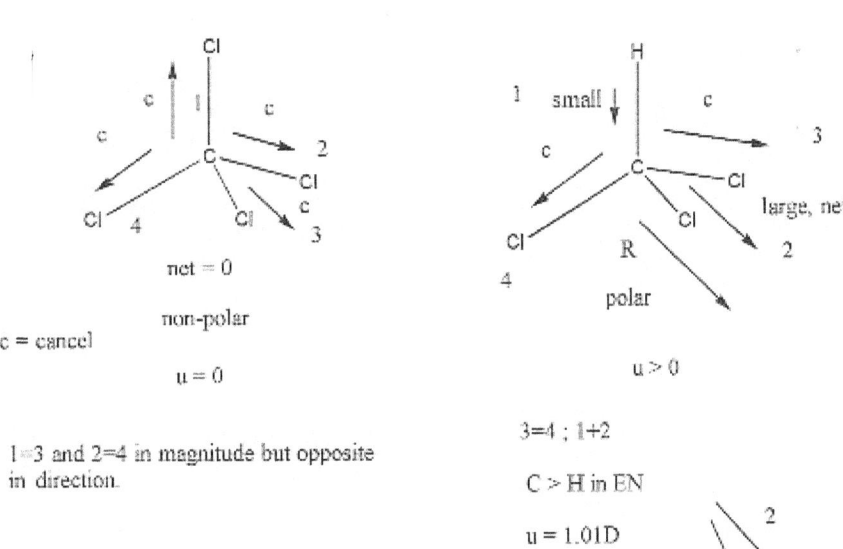

net = 0

non-polar

c = cancel

u = 0

1=3 and 2=4 in magnitude but opposite in direction.

3=4 ; 1+2

C > H in EN

u = 1.01D

polar

u > 0

If we know the coordinates and directions of the vectors, we can use math to find the resultant vector in both magnitude and direction.

Here magnitude can be shown by different lengths of the vectors and the direction by drawing the vector down the bond (360 degrees).

Since all the Dihedral angles are 109*, the resulting vectors will cancel if they are opposite in direction and the same magnitude.

2 u : 2 molecules NOF and NO2F
ul = .47D smaller
u2=1.81D larger

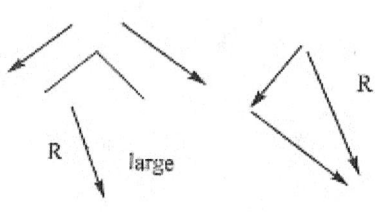

The resultant vectors are pointing in a truer direction.

In an approximate method, the vectors are pointing up or down.

18e

e = TP   1.81D

Mol = angular AX2E

R

R   large

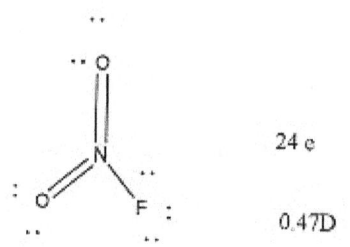

24 e

0.47D

e = TP

Mol = TP AX3

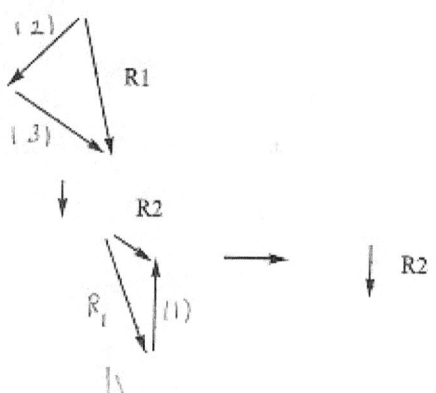

Final R is down and small

R2   small

The higher the EN difference, the larger the magnitude/length of vector.

Additon of vectors can also been done in the following manner which gives the same resultant vector : A+(-B)=C = A-B. -B is opposite direction same magnitude.

The use of vectors in this manner is mainly qualitative. But approximating the length with a little more accuracy and adhering strictly to the orientation/angle, we can be a little more quantitative. The resultant vector direction will be mainly determined by the orientation of the covalent bond. The length/magnitude of the resultant vector will be determined by vector addition.

Valence Bond Theory

Quantum Mechanics uses two more approaches to understanding molecular structures. The first is the Valence Bond Theory and the second one is the Molecular Orbital Theory. The Valence Bond Theory studies the atomic orbitals and how they create a covalent bond. The second theory obviously studies the molecular orbitals created by atomic orbitals.

We have already thoroughly described the Valence Bond Theory in the above sections. Here we will just summarize the highlights of the theory.

A covalent bond is formed by the pairing of 2 e with opposite spin in a region of overlap of atomic orbitals between 2 atoms. The region has a high e charge density. This is the area of maximum probability of finding the e.

Overlap is maximum if they come together Head to Head.: $\rightarrow \leftarrow$. This enables maximum psi function convergence, addition of a wave function.

Hybridization is the excited state of the atomic orbital. This allows more area for overlap and covalent bond formation. By reshaping the orbital, it allows a higher probability for chemical reactivity.

Hybrid Orbitals and Geometry:

1. sp   Linear
2. sp^2 Trigonal Planar
3. sp^3 Tetrahedral
4. sp^3d Trigonal Bipyramidal
5. sp^3d^2 Octahedral

Pentafluoride : IF5

I + F : Group 7A = 7 valence e @
6*7=42 total e
Draw the 5 bonds and then complete the octets. You are left with 2 e = 1 L.P. Which is associated with the central atom, I.
This is 6 orbitals which requires the hybrid = sp^3d^2 = octahedral.

sp^3d^2 hybrid I : e configuration : sp^3d^2   + 5d

                                    ud u u u u u   3 empty

                                    LP

F : p orbital + I : sp^3d^2 hybrid orbital overlap : linear orientation : Hd to Hd.

Bond angle : 90*

Mol geometry : square pyramid/pyramidal : 5 faced pyramid

e geometry : octahedral

VSEPR : AX5E

VBT : sp^3d^2

Each orbit has only 2 e. 1 spin up and 1 spin down. The F atom needs 1 e for an Octet. This leads to stabilization of the atom. The electron configuration notation shows only the valence e. The exponents are the e numbers which add to 7. This is also the group number in the periodic table. The electron geometry is like a flower. The p orbit is obvious and the s orbit is buried. The s orbit is spherical and at the center of the atom. Also it is a 2s type which is bigger and which contains the 1s orbit at it's center.

Notes:

I

Energy

8 Faces

diamond

sp^3d^2
Hybrid

Energy

Orbital

$5s$

$5p$
core omitted

$5s^2 5p^5 = 7e$

sp^3d^2
octahedral
e^- grp

$5d$
valence

molecule grp

F F F F F F

I

bottom face

5 face
pyramid

square

$5d$

sp^3d^2 excited   Hybrid

LP

$5p$

$5o$

$sp^3d^2 = 6$

AX_5E

6 Atomic orbits
6 hybrid orbits

I atom is very large. It too has 7 valence e. The core contains 1-4 subshells! I is the central atom therefore it hybridizes. According to VBT, 6 atomic orbitals generate 6 hybrid orbitals. This is an excitation state for the e in these orbits. The psi wave equation changes to create a larger orbit and a more accessible one. This increases the probability of chemical reaction; therefore covalent bond formation. In math, an easy way to enlarge a wave equation is to increase it's amplitude. For example, 2sin(theta) has a larger amplitude than sin(theta). Graph it.

The hybrid notation's exponents add to 6. Comparing this to VSPER notation, you can see the same number of orbits. Orbits can not created from nothing. Only the psi wave equation has changed. The wave equation is used to represent the area of probability of finding the e. This seems appropriate given the fact that the e can act as a wave particle. This whole hybrid process seeks the minimum potential energy state: the lowest energy state: state of stabilization. So the e in the central atom elevates to a higher energy state to eventually reach the lowest energy state after the covalent bond formation. Prior to the reaction: F and I seeks an e. I hybridizes to create an Octet between I and F. The reaction product is stable. Coulomb's law, electrostatic forces, are the drivers of this dance. The proton nucleus of each atom attracts the e of the other atom : + → ← -.

Multiple bonds : Double and Triple bond

Hybridization can account for the molecular shape and kinds of bonds.

sp^3 has only sigma bonds: end to end type; eg methane

C : atomic orbital : 2s + 2p -> hybrid atomic orbital: 4 sp3 (excited).

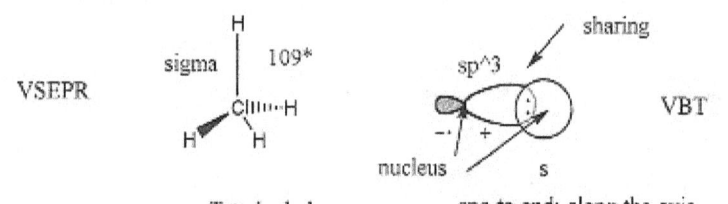

VSEPR

H   109*
sigma |
       C||||···H
H   H

Tetrahedral

sp^3   sharing
              VBT
nucleus   s

one to end: along the axis

H atom: geometric shape of sphere
C atom: geometric shape of ellipse (derivative of sphere)
The volume of sharing is the covalent bond . The origin of the force of attraction is the 2 nuclei. The distance between both nuclei is determined by the balance of the individual forces. The +/- are regions where the wave equation are +/-.
The in phase orbital has the same sign and orientation. When they overlap, they generate a bonding orbita. If they do not, they are anti-bonding orbitals. So you need to characters to create a bonding orbital, 1) same sign + or - and 2) proper orientation. The sigma bond has rotational symmetry (spherical). The pi bond does not, more axial. The pi bond is oriented side by side. Any rotation among the axes, destroys the pi bond.

Ethylene: CH2=CH2

double bond

C atrom: The hybrid, sp^2, of C has an e. config. of: u u u   u
                                                      sp^2   2p

VSEPR can predict multiple bonds: Remember a double bond is 1in bonding.

VSEPR   120*

Trigonal planar
All atoms on the same plane!

pi

VBT can do the same, but also the orientation of the p orbitals that create the pi bond. The p orbitals are oriented perpendicular and side by side.

pi molecular orbital

2 p atomic orbitals -> 2 pi molecular orbitals   unhybridized p orbital

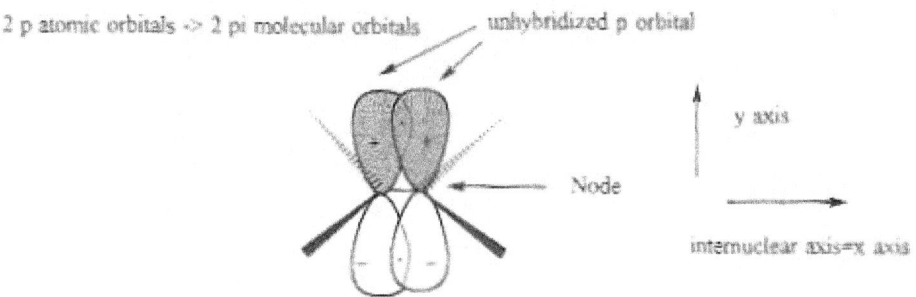

y axis

Node

internuclear axis=x axis

The p orbitals are perpendicular to the internuclear axis.
2 electrons occupy the 2 lobes of the p orbital.
They are in phase : gray to gray and white to white. They have the same sign.
They overlap creating bonding orbits.
If we rotated 1 p orbit around the internuclear axis, this would destroy the bonding orbit.
If they are different signs, they are anti-bonding orbits. These are empty, no e, orbits : *pi.
Node is where the probability density equals 0. The e can not occupy the nucleus position. This node is closest to the nuclei.

Important terms:
Overlap: intersecting volumes of space orbits
bonding: occupying e pair (2e) : q = - : nucleophile
non-bonding: empty : q = + : electrophile
In phase : same sign : convergence of wave equation : summating
Out of phase: opposite sign : cancellation of wave equation

Notes:

sp hybrid=the triple bond

Acetylene

VSEPR   HC≡≡CH   Linear=180*

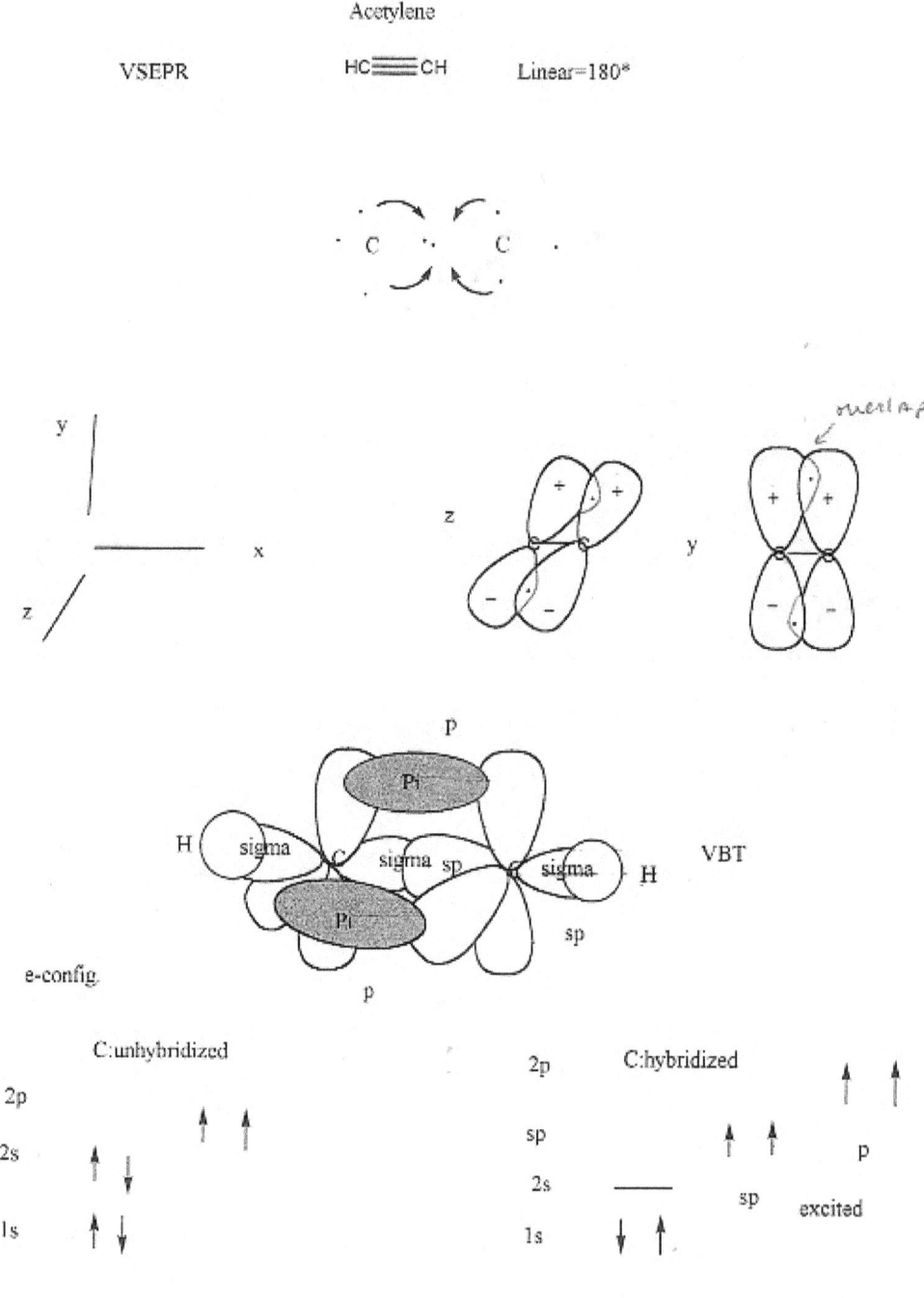

e-config.

C:unhybridized

2p	↑ ↑
2s	↑ ↓
1s	↑ ↓

C:hybridized

VBT

2p   ↑ ↑

sp   ↑ ↑   p

2s   ——   sp

1s   ↓ ↑   excited

Molecular Orbital Theory: MOT

Wave equation (Quantum Mechanics) -> regions in a molecule of high probability of finding e.

Atomic orbitals united/combined to form molecular orabitals 1:1 correspondence.

The nuclei are placed and e(s) are positioned in orbitals that create the lowest potential energy state: stable and favored.

Combining leads to 2 states: 1) high = antibonding orbit *(destroys bonding) 2) low=bonding, sometimes 3) nonbonding (neutral). Molecules prefer a low energy state.

H2: 1s + 1s -> sigma* 1s + sigma 1s

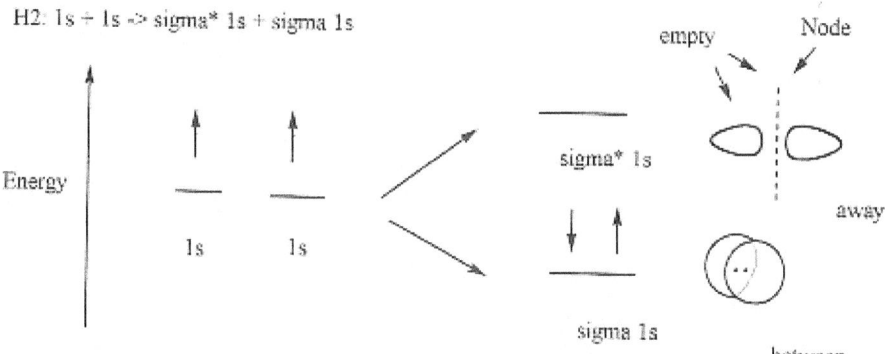

If sigma* was occupied, it would destroy the bond.

The following rules still apply: aufbau, Pauli exclusion, and Hund's.

Bond Order: single, double, triple bond

$$BO = [\text{\#e in bond} - \text{\#e in anti bond}]/2$$

$$H2 = [2-0]/2 = 1$$

H ·· H

H————H

BO=1

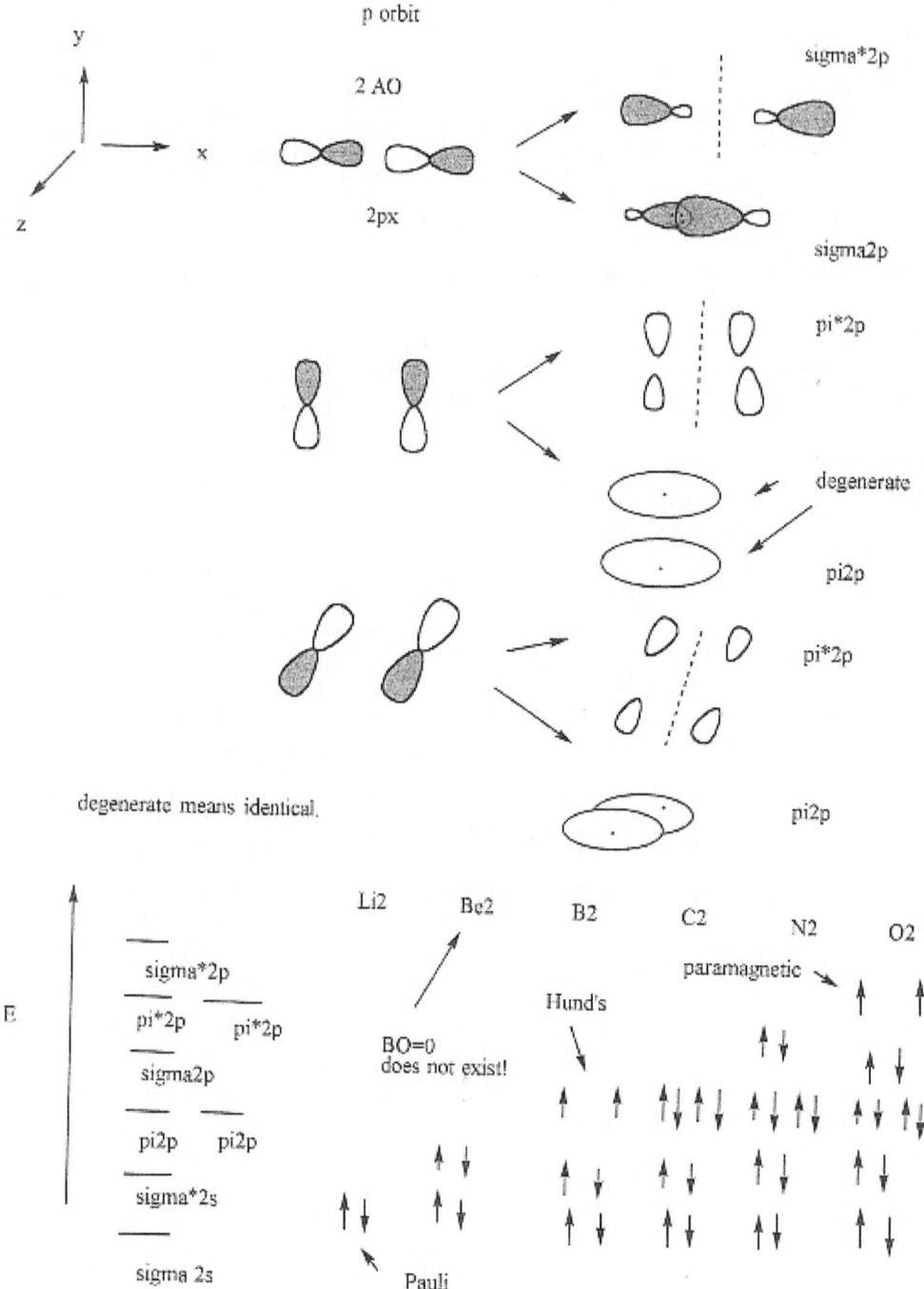

In the sigma2p orbit, the 2 e/s are in the intersecting volume. It is dark. But if you look carefully, you will see 1 e in the volume, the other outside with a arrow pointing inside.

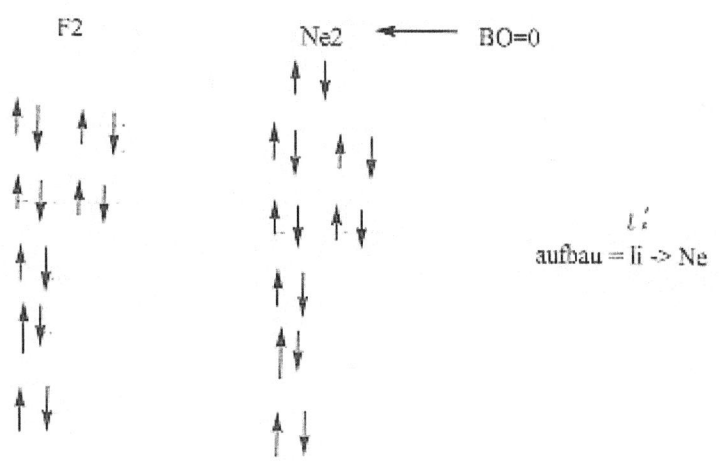

Notes:

Lewis Dot Structures

Atom is stable when it's valence shell is full: lowest potential energy state.

Octet Rule applies to 2nd and 3rd row elements of the periodic table.

Covalent bond is sharing of electrons; 1 e from @ atom. Both atoms seem to gain 1 e.

H ·   He :

Li ·   Be ·   B ·   · C ·   · N ·   · O ··   · F :   ·· Ne ··

Na ·   Mg ·   Al ·   · Si ·   · P ·   · S ··   · Cl ··   ·· Ar ··

Only valence eletrons, no core electrons.

Find the central and peripheral atoms.

If given a formula, draw Lewis structures.

Methane: CH4 : C=central atom, H:peripheral atoms, tetrahedral.

TH

Methanol: CH3OH: C=central, O central, final H to O., terahedral.

Water: O=central, H:peripheral, tetrahedral.

TH   TH

angled

Formaldehyde: CH2O, aldehyde function=CHO, bent, trigonal planar, double bond.

trigonal planar

H——C≡≡≡N ··   linear

Hydrogen Cyanide: HCN: C=central, cyanide function=CN, linear, triple bond.

Polarity

Ionic   ->   Covalent
Polar -> partial* -> non-polar

Polarity signifies partial charge distribution.

Electronegativity is electron attraction
0-4
Upper right hand corner of PT > Lower Left hand corner.
Delta EN: <1 = non-polar
           >1.5 = polar

H
2.2
Li   Be   B   C   N   O   F
.98   1.57 2.04 2.55 3.04 3.44 3.98
Na   Mg   Al   Si   P   S   Cl
.93   1.31 1.61 1.9   2.19 2.58 3.16
K   Ca   Br
.82   1   2.96
                                    I
                                    2.66

Bond dipole: mu=e*d   e=q(charge).   d=distance
                      4.8*10^-10 esu (electrostatic units)   10^-8 cm
Debye unit (D) = 1*10^-18 esu*cm

H —— Cl

Dipole moment = vector sum of all dipole bonds

The greater the polarity, the less stable, and more reative.

Bond	Bond Dipole, D	Bond	Bond Dipole, D
C—F	1.53	H—F	1.82
C—Cl	1.59	H—Cl	1.08
C—Br	1.48	H—Br	0.82
C—I	1.29	H—I	0.44
C—N	0.22	H—N	1.32
C—O	0.85	H—O	1.53
C—H	0.35		

C, N, O, and F are close is the periodic table. H and F are on opposite sides of the table. H has a large dipole.

Notes:

Induction vs. Field Effects

Induction is creating increase EN in an adjacent atom through a bond.

Field effect is the same, except it is done through space, proximity.

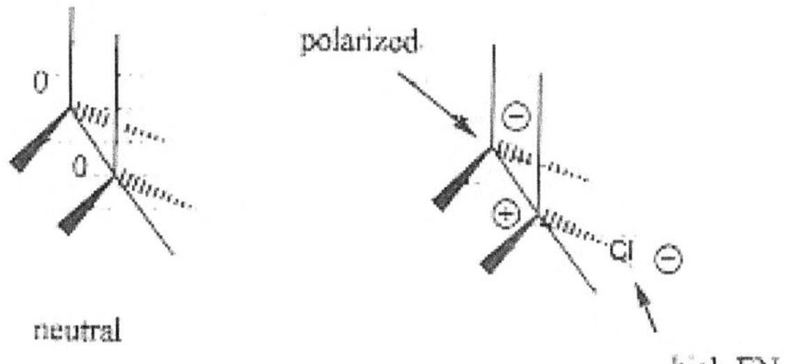

neutral

Induction

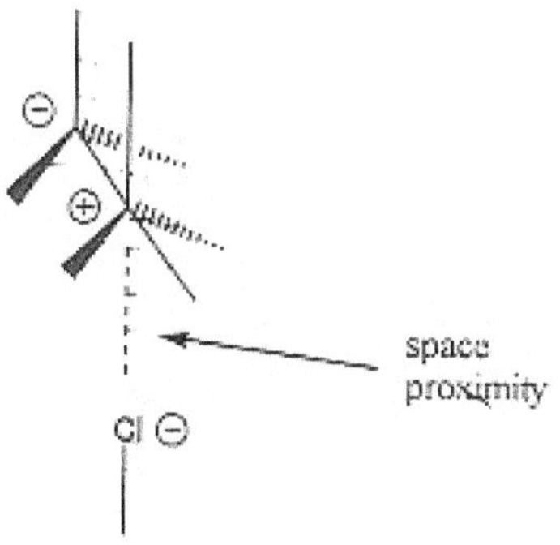

Field Effect . . .

Formal Charge

Ions

FC = # valence e - # non-bonding e - 1/2 # bonding e

Methane

H: 1-0-1/2*0=0

C:4-0-1/2*8=0

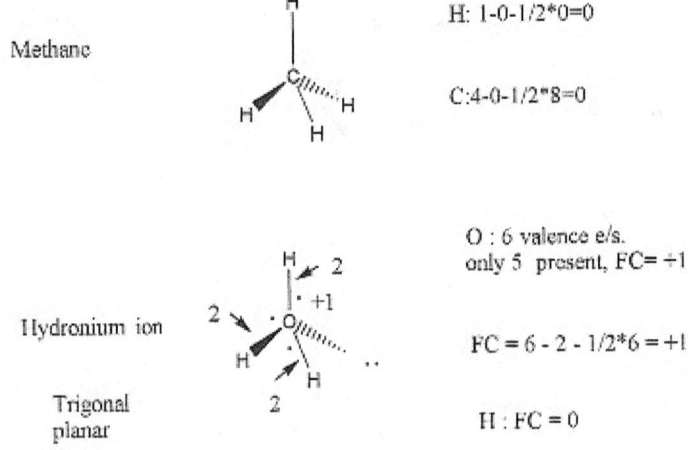

Hydronium ion

Trigonal
planar

O : 6 valence e/s.
only 5 present, FC= +1

FC = 6 - 2 - 1/2*6 = +1

H : FC = 0

FC is based on Lewis structure.

Methoxide ion : CH3O^(-1)

Tetrahedron

O: 6 valenc e/s.
7 present,FC= -1

6 - 6 - 1/2*2 = -1

Resonance

One Lewis structure does not describe the molecule correctly.

The net charge is throughout the entire molecule.

The resonance hybrid is the actural structure. It is a weighted average of the contributors.

The other structures are resonance contributors, which can be major or minor types.

Separation of charge expends energy and becomes a minor contributor.

Rules
1 non-bonding and pi-bonding e's change position.
2 Atoms do not change position.
3 All contributors have same # of paired and unpaired e's.
4 Contributors from 2nd period with 8 e's are more important.
5 Contributors with greater # of covalent bonds are more important.
6 P & S have empty 3d orbitals, can write 10 or more e's.
7 No separation of charge is more important.
8 (-) q on most EN atom or (+) q on least EN atom is important.

Carbonate ion

nitroethene

Separation of q

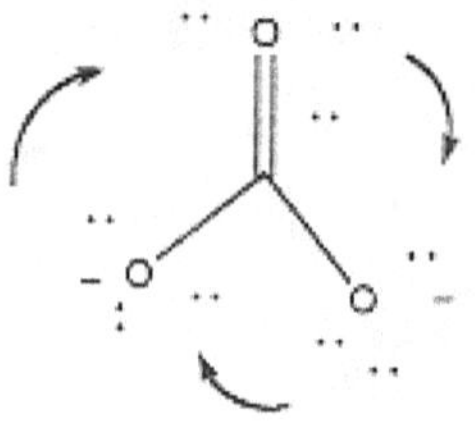

Double bond can float around the entire molecule.

e pair and adjacent double bond is perfect for resonance.

The q distributes around the entire molecule.

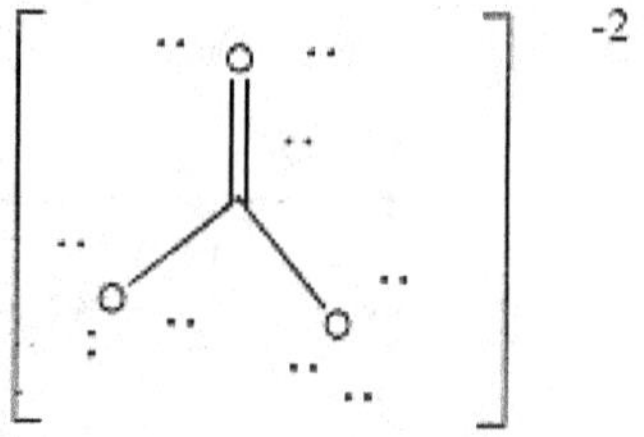

resonance hybrid is the structure.

All resonance contributors are major.

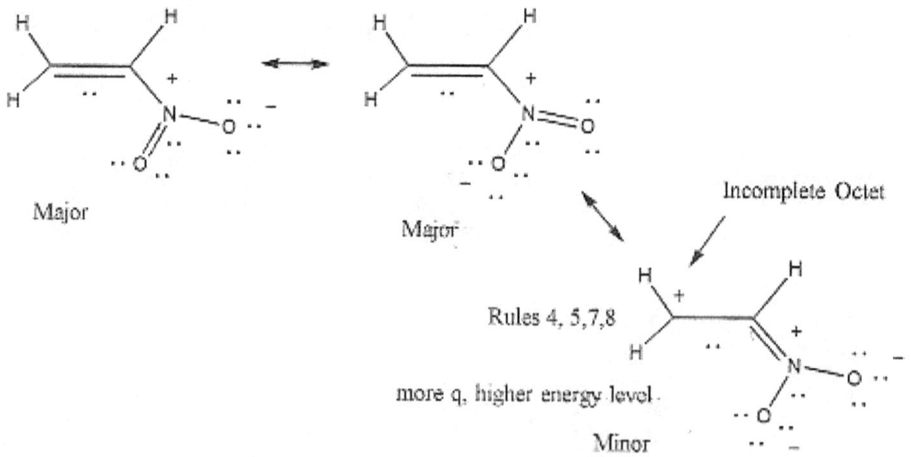

Major

Major

Rules 4, 5,7,8

more q, higher energy level.

Incomplete Octet

Minor

Resonance

One Lewis structure does not describe the molecule correctly.

The net charge is throughout the entire molecule.

The resonance hybrid is the actural structure. It is a weighted average of the contributors

The other structures are resonance contributors, which can be major or minor types.

Separation of charge expends energy and becomes a minor contributor.

Rules:
1. non-bonding and pi bonding e/s change position.
2. Atoms do not change position.
3. All contributors have same # of paired and unpaired e/s.
4. Contributors from 2nd period with 8 e/s are more important.
5. Contributors with greater # of covalent bonds are more important.
6. P & S have empty 3d orbitals, can write 10 or more e/s.
7. No separation of charge is more important.
8. (-) q on most EN atom or (+) q on least EN atom is important.

Carbonate ion

nitroethene

Separation of q